FLEXIBLE CIRCUITS

ELECTRICAL ENGINEERING AND ELECTRONICS

A Series of Reference Books and Textbooks

Editors

Marlin O. Thurston
Department of Electrical
Engineering
The Ohio State University
Columbus, Ohio

William Middendorf
Department of Electrical
and Computer Engineering
University of Cincinnati
Cincinnati, Ohio

1. Rational Fault Analysis, *edited by Richard Saeks and S. R. Liberty*
2. Nonparametric Methods in Communications, *edited by P. Papantoni-Kazakos and Dimitri Kazakos*
3. Interactive Pattern Recognition, *Yi-tzuu Chien*
4. Solid-State Electronics, *Lawrence E. Murr*
5. Electronic, Magnetic, and Thermal Properties of Solid Materials, *Klaus Schröder*
6. Magnetic-Bubble Memory Technology, *Hsu Chang*
7. Transformer and Inductor Design Handbook, *Colonel Wm. T. McLyman*
8. Electromagnetics: Classical and Modern Theory and Applications, *Samuel Seely and Alexander D. Poularikas*
9. One-Dimensional Digital Signal Processing, *Chi-Tsong Chen*
10. Interconnected Dynamical Systems, *Raymond A. DeCarlo and Richard Saeks*
11. Modern Digital Control Systems, *Raymond G. Jacquot*
12. Hybrid Circuit Design and Manufacture, *Roydn D. Jones*
13. Magnetic Core Selection for Tranformers and Inductors: A User's Guide to Practice and Specification, *Colonel Wm. T. McLyman*
14. Static and Rotating Electromagnetic Devices, *Richard H. Engelmann*
15. Energy-Efficient Electric Motors: Selection and Application, *John C. Andreas*
16. Electromagnetic Compossibility, *Heinz M. Schlicke*
17. Electronics: Models, Analysis, and Systems, *James G. Gottling*

Other Volumes in Preparation

FLEXIBLE CIRCUITS
Design and Applications

STEVE GURLEY
Rogers Corporation
Chandler, Arizona

With contributions by

CARL A. EDSTROM, JR.
RAY D. GREENWAY
WILLIAM P. KELLY

MARCEL DEKKER, INC.   New York and Basel

Library of Congress Cataloging in Publication Data

Gurley, Steve, 1936-
  Flexible circuits.

  (Electrical engineering and electronics; 20)
  Includes index.
  1. Flexible printed circuits. I. Title. II. Series.
TK 7868.P7G87   1984   621.381'74   84-5032
ISBN 0-8247-7215-6

MARCEL DEKKER, INC.
270 Madison Avenue, New York, New York 10016

Current printing (last digit):
10 9 8 7 6 5 4 3 2 1

PRINTED IN THE UNITED STATES OF AMERICA

Preface

This book was written to provide basic guidelines relative to the design, application and manufacturing considerations of single-sided and double-sided, plated-through hole flexible circuits which are commonly used in both static and dynamic applications.

The book will help solve problems involving packaging and costs in multiplant interconnection situations. The objective of the book is to inform packaging, electrical, and mechanical engineers of the most economical choices relative to the design and specification of the flexible circuit interconnect. Another objective is to provide sufficient information concerning the interconnection of electrical/electronic systems, so the lowest cost systems approach can be easily chosen.

One element of the flexible circuit market that has not been covered in depth is flexible multilayer and rigid-flex multilayer circuits. The reason for this is twofold. First, most multilayer and rigid-flex applications are manufactured in small quantities and are for defense or military applications requiring a completely different set of guidelines. Second, flexible multilayer circuits are not really flexible but flexible materials are used for space and weight reduction. When many layers are fabricated from a group of flexible films, adhesives, and foils, they become a fairly rigid piece of material when all laminated together.

Steve Gurley

Contents

1
Description and Construction of Flexible Circuits

1.1. DESCRIPTION OF FLEXIBLE CIRCUITS

1.1.1 IPC Definition

Flexible printed wiring, sometimes called flexible printed circuitry can be defined as a random arrangement of printed wiring, utilizing flexible base material with or without cover layers. This definition, along with many other printed circuit terms and definitions can be found in the publication *ANSI/IPC-T-50B*, published by the Institute for Interconnection and Packaging Electronic Circuits, revised in June 1980.

1.1.2 Identification Features

Identification features can be easily drawn from the definition of the product. It is necessary to develop an almost automatic recognition of flexible circuit characteristics since there are relatively few manufacturers of this product. Much confusion occurs when manufacturers of similar products are solicited to supply flexible circuits when in fact, they don't manufacture them.

1.1.3 Interconnect Look-alikes

Interconnection systems which are similar to, and confused with, flexible circuits are called flat cable, collated cable, ribbon cable, and sometimes wiring harnesses. The basic differences among all of these interconnection products are the types and forms of conductors, and the types and forms of insulating materials used. The most significant difference between flexible circuitry and "look-alikes" is the random arrangement of conductors, that is that all conductors are not parallel to each other. This factor is generally dictated by the arrangement of electrical/electronic components which are to be interconnected on one or more position planes.

Interconnection look-alikes, which sometimes appear to be flexible circuits but which are really wiring harnesses, woven cables, or other types of insulated flat or round conductor cables are shown in Figure 1.1.

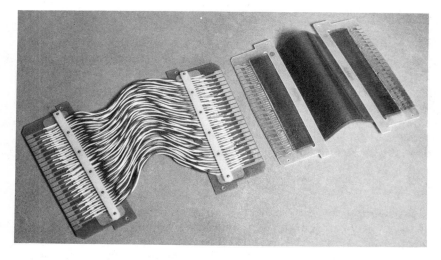

Figure 1.1 Interconnection look-alikes. Flexible cable and flat cable, which sometimes look like flexible circuitry. (Courtesy of Rogers Corporation, Chandler, Ariz.)

1.2 CONSTRUCTION OF FLEXIBLE CIRCUITS

The construction of flexible circuits is generally thought of in
two- or three-component part systems. In the two-part system, base insulator materials are joined to the conductor foil
without an adhesive system, either by additive processes such
as electroless and electrolytic plating or by mechanically screening an image onto the insulator surface with conductive epoxy.
Another two-part system is to use resins in the base insulator
system to bond the copper in the original laminating cycle.
The three-part system is composed of a separate adhesive system which is usually coated onto the base insulating film and
then nip-rolled together with the conducting foil to produce
the laminate. Figures 1.2 and 1.3 show screening and laminating operations, respectively. Figure 1.4 shows a typical
roll of finished laminate.

1.2.1 Films and Other Base Insulations

From a visual standpoint, flexible circuits look very much like
printed circuit hardboards unless they are viewed after being
formed into a multiplane interconnect system. Some of the confusion surrounding the design and application of flexible circuits seems to be based on the fact that flexible circuits in a
single flat plane have an appearance often mistaken for their
hardboard cousins.

   The major difference between the two products is that flexible circuits are usually fabricated using conductive foils of
extremely ductile copper, which are glued to extremely thin,
flexible substrates with a number of adhesive choices. Most of
these thin substrates are in the film family, although some are
base insulators manufactured using a combination of strength-
ening fiber, in either a random or mat form held together with
a resin. In such cases the resin itself acts as the adhesive
for the conductive foil layer and the product is manufactured
in a two-part form rather than the more common three-part
base adhesive and conducting foil configuration.

Polymide

The most popular insulating base material is the polyimide film,
popularized in the electronics industry by E. I. du Pont de

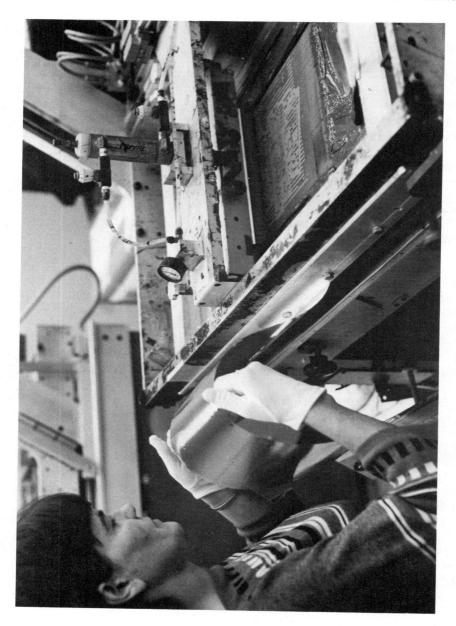

Figure 1.2 An example of the screening process. (Courtesy of Sheldahl, Inc., Northfield, Minn.)

Figure 1.3 An example of the laminating process. (Courtesy of Sheldahl, Inc., Northfield, Minn.)

Figure 1.4 A typical roll of finished laminate for flexible circuits. (Courtesy of Rogers Corporation, Chandler, Ariz.)

Nemours and Co. and manufactured under the trade name of Kapton.

According to Du Pont, Kapton® polyimide film possesses a unique combination of properties previously unavailable in circuit materials. The ability of Kapton to maintain excellent physical, electrical, and mechanical properties over a wide temperature range has proven especially useful when used in flexible circuit applications involving high temperatures, such as those created in soldering applications. Kapton is synthesized by a polycondensation reaction between an airomatic and dianhydride and airomatic diamine. There is no known organic solvent for this film and it is infusible and flame-resistant. Besides being used for flexible circuits, Kapton film is used in other electrical and electronic applications, such as insulating wire and cable, as slot liners in motors, and for transformer insulation.

Although Kapton is available in several different types, the type H material is the most popular for flexible circuit use and is provided in thicknesses of 0.001, 0.002, 0.003, and 0.005 inches. It is also available on special order in 0.0005 inches for circuit applications requiring extreme thinness and flexibility. When the best dimensional stability is needed, type V Kapton is available, however, it is available only in 0.002, 0.003, and 0.005 in. thicknesses. Table 1.1 and Figures 1.5 a-d show the physical and chemical properties which give this film its outstanding characteristics.

Table 1.1 Typical Properties of Kapton Type H Film 25 μm (1 mil)

PHYSICAL	Typical Values			Test Method
	78K (195°C)	296K (23°C)	473K (200°C)	
Ultimate (MD) Tensile Strength, MPa (psi)	241 (35,000)	172 (25,000)	117 (17,000)	ASTM D-882-64T
Ultimate (MD) Elongation	2%	70%	90%	ASTM D-882-64T
Tensile Modulus, GPa (MD) (psi)	3.5 (510,000)	3.0 (430,000)	1.86 (260,000)	ASTM D-882-64T
Tear Strength — Propagating (Elmendorf), g	—	8	—	ASTM D-1922-61T
Tear Strength — Initial (Graves), g(g/mil)	—	510 (510)	—	ASTM D-1004-61
MD — Machine Direction				

THERMAL	Typical Values	Test Condition	Test Method
Zero Strength Temperature	1088K (815°C)	.14MPa (20 psi) load for 5 seconds	Du Pont Hot Bar Test
Coefficient of Linear Expansion	2.0 × 10⁻⁵m /m /K (2.0 × 10⁻⁵in./in./°C)	259 to 311K (—14°C to 38°C)	ASTM D-696-44
Flammability	94 VTM-O		UL-94 (1-24-80)
Limiting Oxygen Index	100H-38		ASTM D-2863-74

ELECTRICAL	Typical Value	Test Condition	Test Method
Dielectric Strength 25 μm (1 mil)	276 v/μm (7,000 v/mil)	60 hertz 1/4" electrodes	ASTM D-149-61
Dielectric constant 25 μm (1 mil)	3.5	1 kilohertz	ASTM D-150-59T
Dissipation Factor 25 μm (1 mil)	.0025	1 kilohertz	ASTM D-150-59T

CHEMICAL		
Chemical resistance	Excellent (except for strong bases)	
Moisture Absorption 25 μm (1 mil)	1.3% Type H 2.9% Type H & V	50% Relative Humidity at 296K (23°C) Immersion for 24 hours at 296K (23°C)

Source: Courtesy of E. I. du Pont de Nemours and Co., Wilmington, Del.

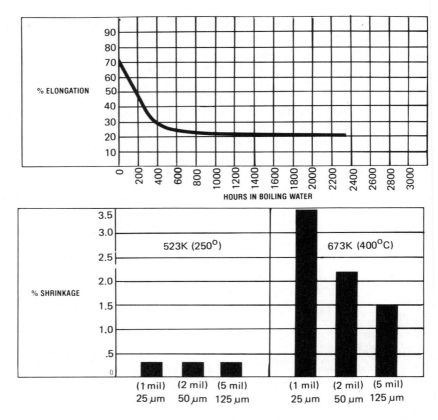

Figure 1.5a Top: Ultimate elongation after exposure in 373K
(110°C) water: Type H 25 μm (1 mil) Kapton polyimide film.
Bottom: Residual shrinkage versus exposure, temperature,
and gauge at 523K (250°C) and 673K (400°C) for 1, 2, and 5
mil Kapton® polyimide film. (Courtesy of E. I. du Pont de
Nemours & Co., Polymer Products Division, Wilmington, Del.)

Polyester

Polyester film is very popular for use in the manufacture of
flexible circuits, owing to its electrical characteristics and, because, at the present time, it is the lowest cost popular film
used in flexible circuit fabrication. Polyester film is made
from polyester terephthalate, which is a polymer formed by

Temperature Range	(K) m/m·K x 10^{-5}
296-373K (23-100°C)	1.80
373-473K (100-200°C)	3.10
473-573K (200-300°C)	4.85
573-673K (300-400°C)	7.75
296-673K (23-400°C)	4.55

Figure 1.5b Thermal coefficient of expansion, Type H film, 25 μm (1 mil), thermally exposed. (Courtesy of E. I. du Pont de Nemours & Co., Polymer Products Division, Wilmington, Del.)

the condensation reaction of ethylene glycol and terephalic acid. This film has also been widely promoted by E. I. du Pont de Nemours and Co. Inc. under the trade name Mylar. Mylar, unlike other low-cost films, does not contain any plasticizers, and therefore, does not become brittle in normal conditions. It has a high dielectric strength of 7.5 kV per mil when measured on 0.001 in. film. It has a tensile strength of 25,000 psi and is very tolerant or resistant to most chemicals and moisture (Table 1.2).

It is used primarily in low-cost consumer applications where low cost combines well with high-volume usage. This film can withstand a temperature range from -70 to +150°C. The relatively low level of temperature capability is the main limiting factor of this film. Its use, therefore, in flexible circuit applications requiring high-temperature resistance such as is needed for soldering operations is extremely limited and only in those cases where a high degree of engineering and processing sophistication is available can it be used in a satisfactory way. Mylar is available in a wide range of thicknesses, however, a half mil thru 0.005-in. thicknesses are the most popular for flexible circuit applications. The accompanying

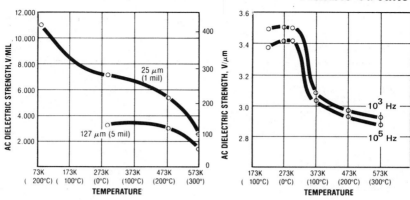

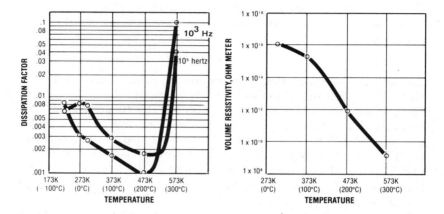

Figure 1.5c Effects of temperature on Kapton polyimide film.
(Courtesy of E. I. du Pont de Nemours & Co., Polymer Products Division, Wilmington, Del.)

charts show the physical and chemical properties which gives
this film the characteristics desired for many low-cost flexible
circuit applications.

Random Fiber Aramids

Random fiber material, as well as insulating materials in mats
and impregnated with resins, are often used in specialized

applications. A random fiber aramid material, sold under the trade name Nomex is sometimes used when high-temperature resistance is desired for applications, but where low cost is still essential. Nomex is manufactured by du Pont and is used to meet many insulating requirements where high-temperature resistance is important but where some processing deficiencies

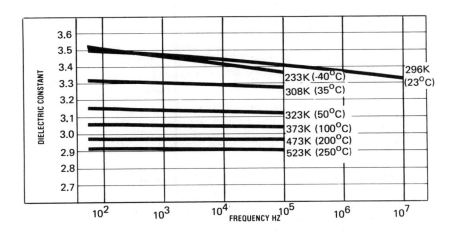

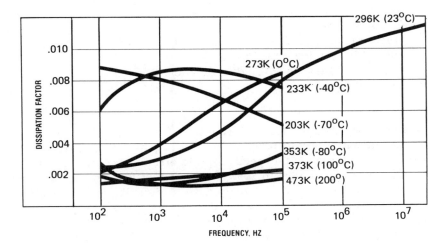

Figure 1.5d Effect of frequency on Kapton polyimide Type H film, 25 μm (1 mil).

Table 1.2 Typical Property Values of Mylar[a] Polyester Film Type EL for Printed Circuits

THERMAL	
Melting Point	Approximately 250°C (480°F) (Fisher Johns Method)
Coefficient of Thermal Expansion	1.7 x 10⁻⁵ in./in./°C (30°C-50°C, ASTM D 696-44 modified)
Service Temperature	−70°C to 150°C (−100°F to 300°F) (Soldering of circuits on "Mylar" can be done at solder temperatures up to about 275°C (530°F)
Strain Relief	1.5% (30 minutes at 150°C [300°F]b
PHYSICAL	
Tensile Strength	25,000 psi (25°C, ASTM D 882-64T Method A)
Elongation	120% (25°C, ASTM D 882-64T Method A)
Tensile Modulus	550,000 psi (25°C, ASTM D 882-64T Method A)
Density	1.395 (25°C, ASTM D 1505-63T modified)
CHEMICAL	
Moisture Absorption	Less than 0.8% (ASTM D 570-63; 24 hour immersion at 23°C)
ELECTRICAL	
Dielectric Strength (1-mil)	7500 volts/mil (25°C, 60 Hz, ASTM D 149-64)
Dielectric Strength (1-mil)	5000 volts/mil (150°C, 60 Hz, ASTM D 149-64)
Dielectric Constant	3.0-3.7 (25°C-150°C, 60 Hz-1 MHz, ASTM D 150-65T)
Dissipation Factor	0.005 (25°C, 1 KHz, ASTM D 150-65T)
Volume Resistivity	10¹⁸ ohm-cm (25°C, ASTM D 257-66) 10¹³ ohm-cm (150°C, ASTM D 257-66)

[a]Reg. U.S. Pat. Off.

[b]Typical strain relief for experimental low shrink film is 1.0% at 150°C (300°F) for 30 minutes.

Source: Courtesy of E. I. du Pont de Nemours & Co., Wilmington, Del.

do not preclude its use. Weaknesses of this material include its low initial tear and propagation strengths. With careful handling of the manufactured flexible circuit and with the inclusion of antitear design features, such as film and conductor stops, this product can be used with a high degree of reliability. Other random and glass-mat-base materials use combinations of nylon, dacron, and polyester fibers in a matte and/or random formation to provide insulating materials which have many fine characteristics. Examples of the physical and chemical characteristics of these random fiber materials are shown on the following pages (Tables 1.3, 1.4, Figures 1.6, 1.7).

Table 1.3 Typical Physical Properties[a] of Nomex Aramid Paper Type 410 (MD = Machine Direction, XD = Cross Direction of Paper)

Nominal Thickness	mil		2	3	5	7	10	12	15	20	24	29	30
	(mm)		(0.05)	(0.08)	(0.13)	(0.18)	(0.25)	(0.30)	(0.38)	(0.51)	(0.61)	(0.74)	(0.76)
Basis Weight	oz/yd²		1.2	1.9	3.4	5.1	7.3	9.1	11	16	20	25	25
	(g/m²)		(40)	(63)	(120)	(170)	(250)	(310)	(370)	(540)	(680)	(850)	(850)
Tensile Strength (ASTM[b] D-828-60)	lb/in.	MD	22	40	83	130	180	230	280	350	460	540	510
		XD	11	19	39	64	90	110	150	230	310	370	350
	(N/cm)	MD	(39)	(70)	(150)	(230)	(320)	(400)	(490)	(610)	(810)	(950)	(890)
		XD	(19)	(33)	(68)	(110)	(160)	(190)	(260)	(400)	(540)	(650)	(610)
Elongation	%	MD	9	12	16	20	21	21	21	21	19	17	20
		XD	7	9	13	16	18	18	18	18	14	13	16
Finch Edge Tear (ASTM D-827-47)	lb	MD	21	41	85	130	180	200	220	210	170	200	270
		XD	9	16	33	58	71	86	89	100	79	86	130
	(N)	MD	(93)	(180)	(380)	(580)	(800)	(890)	(980)	(930)	(760)	(890)	(1200)
		XD	(40)	(71)	(150)	(260)	(320)	(380)	(400)	(450)	(350)	(380)	(580)
Elmendorf Tear (ASTM D-689)	g	MD	84	120	220	340	560	690	890	1200	—	—	—
		XD	160	250	580	750	1000	1200	1500	1900	—	—	—
	(N)	MD	(0.8)	(1.2)	(2.2)	(3.3)	(5.5)	(6.8)	(8.7)	(12)	—	—	—
		XD	(1.6)	(2.5)	(5.7)	(7.4)	(9.8)	(12)	(15)	(19)	—	—	—
Shrinkage at 300°C	%	MD	1.7	1.1	0.6	0.5	0.5	0.3	0.3	0.3	0.3	0.3	0.3
		XD	1.3	0.9	0.5	0.4	0.4	0.3	0.3	0.3	0.3	0.3	0.3
Limiting Oxygen Index (ASTM D-2863-70)								0.24 – 0.28						

[a] Not to be used for product specification purposes.
[b] American Society for Testing and Materials, Philadelphia, PA.
Source: Courtesy of E. I. du Pont de Nemours & Co., Wilmington, Del.

Table 1.4 Typical Electrical Properties[a] of Nomex Aramid Paper Type 410 (Measured at 23°C, 50% RH)

Nominal Thickness	mil	2	3	5	7	10	12	15	20	24	29	30
	(mm)	(0.05)	(0.08)	(0.13)	(0.18)	(0.25)	(0.30)	(0.38)	(0.51)	(0.61)	(0.74)	(0.76)
Dielectric Strength												
AC Rapid Rise (ASTM D-149[b])	V/mil	500	610	730	900	880	880	860	780	750	700	660
	(kV/mm)	(20)	(24)	(29)	(35)	(35)	(35)	(34)	(31)	(30)	(28)	(26)
Dielectric Constant (ASTM D-150[c])	60 Hz	1.6	1.6	2.4	2.7	2.7	2.9	3.2	3.4	3.7	3.7	3.7
Dissipation Factor (ASTM D-150[c])	60 Hz	0.004	0.005	0.006	0.006	0.006	0.007	0.007	0.007	0.007	0.007	0.007

[a]Not to be used for product specification purposes.
[b]2-in (51-mm) diameter electrodes.
[c]1-in (25-mm) diameter electrodes under 20 lb/in^2 (140 kPa) pressure.
Source: Courtesy of E. I. du Pont de Nemours & Co., Wilmington, Del.

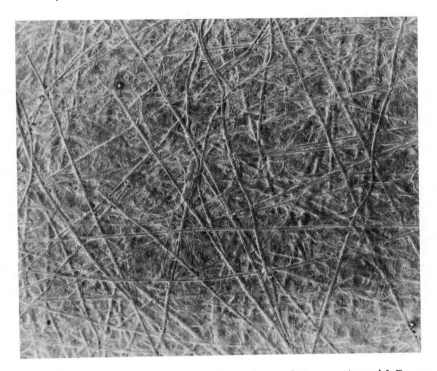

Figure 1.6 Photomicrograph of surface of Nomex Aramid Paper Type 410. (Courtesy of E. I. du Pont de Nemours & Co., Polymer Products Division, Wilmington, Del.)

Other Base Insulations

There are other base insulating materials which are occasionally used for very specialized applications. These materials are sometimes selected because of the application demands, however, there are very few products which offer as broad a range of application capabilities as the types mentioned earlier in this chapter. There are some materials which are very unique and have been developed for use by specific industries. One of these is an epoxy dacron or an epoxy dacron glass, in which man-made fibers such as glass, polyester, and dacron are formed into nonwoven mats. These nonwoven mats are impregnated with epoxy to form the base substrate. At the time

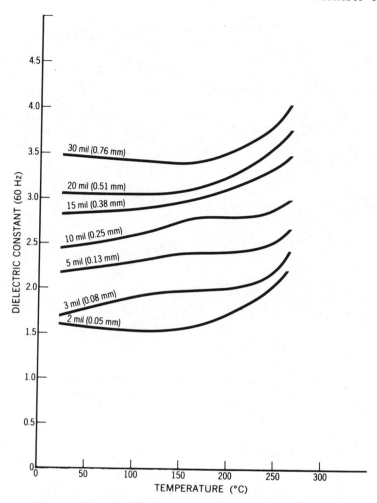

Figure 1.7 Dielectric constant vs. temperature for Nomex aramid paper type 410. (Courtesy of E. I. du Pont de Nemours & Co., Polymer Products Division, Wilmington, Del.)

of manufacture, copper foil is added to one or both sides of the material and the resultant laminate becomes a two-part system because of the absence of a specific adhesive which ordinarily would attach the copper to the base material. In some very-high-temperature applications there are materials which are used such as polyamide-imide TFE Teflon and FEP Teflon. Some of these high-temperature systems have characteristics which are undesirable in applications requiring high flexibility and precise positioning of conductors.

In the low-temperature area, polyvinyl chloride (PVC) is sometimes used when low cost is paramount, however, in most low-cost applications polyester film has processing properties which make it a much more superior product for economical flexible circuitry. The best rule-of-thumb to follow is to contact appropriate application engineers from one or more flexible circuit manufacturers. These application specialists can recommend the best materials to use for any given circuit application, and will generally offer a selection which involves the fewest design compromises combined with the lowest possible cost. It is important to make this kind of engineering contact during the early stage of circuit development.

1.2.2 Foils

There are many foils used as conductors for flexible printed circuitry. Foils include copper, aluminum, nickel, beryllium copper, nichrome, and inconel. The most popular conductor material is made from copper. Copper has been used since the early 1940s, subsequent to the development of printed circuitry. Because copper was used in the building trade prior to its use in the printed circuit industry, it is quantatively measured in ounces per square foot. The shortened definitions are 1 oz, 2 oz, 3 oz, and so on. One ounce copper is 0.0014 in. thick, therefore, 2 oz copper would be 0.028 in. thick, and so on. Copper is available in two basic types: electrodeposited and rolled copper.

Electrodeposited Copper

Electrodeposited copper is manufactured in an electrolytic-plating tank, where the copper from anodes is plated onto a stainless steel drum in varying thicknesses (Fig. 1.8). The longer the plating action continues, the thicker the copper

Figure 1.8 Electrodeposited copper production line. (Courtesy of Gould, Inc., Foil Division, East Lake, Ohio.)

Figure 1.9 Micrograph of grain structure of electrodeposited copper. (Courtesy of Gould, Inc., Foil Division, East Lake, Ohio.)

Table 1.5 Mechanical Properties

CLASS DESCRIPTION

1	Standard electrodeposited	
2	High ductility electro-deposited (flexible applications with minor strain)	
3	High temperature elongation electrodeposited (see special products section) (high reliability multilayer applications where severe stresses are imposed)	
4	Annealed electrodeposited (high reliability multilayer and flexible applications where severe strains are imposed without high tensile requirements)	

Minimum Values

Electrodeposited Copper Class	Copper Weight oz	At room temperature 23° C				At elevated temperature 180° C			
		Tensile		% Elongation (2.0″ GL)		Tensile		% Elongation (2.0″ GL)	
		lb/in²	kg/mm²	CHS 2″/M		lb/in²	kg/mm²	CHS 0.05″/M	
1	½	15,000	10.55	2.0		Not applicable			
	1	30,000	21.09	3.0					
	2+	30,000	21.09	3.0					
2	½	15,000	10.55	5.0		Not applicable			
	1	30,000	21.09	10.0					
	2+	30,000	21.09	15.0					
3	½	15,000	10.55	2.0		--	--	--	
	1	30,000	21.09	15.0		20,000	14.09	2.0	
	2+	30,000	21.09	22.0		25,000	17.62	3.0	
4	1	20,000	14.09	10.0		15,000	10.55	4.0	

Foil Specifications

Thickness by Weight		Thickness by Gauges	
oz/ft²	g/m²	Nom. Inch	Nom. Millimeter
⅛	44	—	0.005
¼	80	—	0.009
⅜	106	—	0.012
½	153	0.0007	0.018
1	305	0.0013	0.035
2	610	0.0028	0.071
3	915	0.0042	0.106
4	1221	0.0056	0.143
5	1526	0.0070	0.178
6	1831	0.0084	0.213
7	2136	0.0098	0.246
10	3052	0.0140	0.353
14	4272	0.0196	0.492

Source: Courtesy of Gould, Inc., Foil Division, East Lake, Ohio.

foil buildup is on the drum. When the copper is removed from the drum as a finished foil it has a very smooth surface on one side and a fairly rough surface on the plating or outside surface.

The advantage of this process is that it produces a foil which has a "tooth" providing a highly satisfactory surface for copper adhesion to various plastic films using any high-grade adhesive system.

As can be seen in Figure 1.9, electrodeposited copper has a grain structure which is vertical, and although this provides a good adhesive surface, it does not exhibit characteristics which lend it to applications where high flexibility is required. There are various treatments available which can be specified by the user which enhance both flexibility and adhesion characteristics. Typical copper treatments and characteristics are shown in Table 1.5.

Rolled Annealed Copper (Figs. 1.10a-d—1.15)

Rolled annealed copper is manufactured by melting cathode copper, which is produced electrolytically, and then casting this copper into ingots using the direct chill method. This particular casting method allows a controlled solidification for close continuous purity monitoring and grain size selection and also eliminates existing defects, such as voids, that would influence the quality of the product when rolled into its final foil form.

The copper ingots are fairly large, weighing over 9 tons. They are hot-rolled to an intermediate gage and milled on all surfaces to insure freedom from defects. After this milling operation the copper is cold-rolled and annealed to the manufacturers specifications and then processed in a specially designed

COMPOSITION LIMITS:	Copper plus Silver 99.9% Min. Oxygen .03% Max	
APPLICABLE SPECIFICATIONS:	ASTM B-451 Copper 110 MIL-F-55561 IPC-CF-150	
PHYSICAL PROPERTIES:	English Units	Metric Units
Melting Point	1981° F	1083° C
Density	.322 lb/in^3	8.91 g/cm
Coefficient of Thermal Expansion	.0000098/°F (68-572° F)	.0000177/°C (20-300° C)
Thermal Conductivity (Annealed)	226 Btu ft/ ft hr °F @ 68° F	.934 cal-cm/cm sec-°C @ 20° C
Electrical Resistivity (Annealed)	10.3 ohm circ mil/ft @ 68° F	1.71 $\mu\Omega$ cm @ 20° C (volume) .152361 ohm g/m^2 (weight)
Electrical Conductivity (Annealed)	101% I.A.C.S.[a] @ 68° F	.586 mΩ /cm @ 20° C
Thermal Capacity (Specific Heat)	.092 Btu/lb/°F @ 68° F	.092 cal/g/°C @ 20° C
Modulus of Elasticity (Tension)	17,000,000 psi	12,000 kg/mm
Modulus of Rigidity	6,400,000 psi	4,500 kg/mm

[a]International Annealed Copper Standard

Figure 1.10a Specifications and physical properties of rolled annealed copper alloy 110. (Courtesy of Somers Thin Strip Division, Olin Brass, Olin Corporation, Waterbury, Conn.)

ANNEALED AND AS ROLLED TEMPERS

Temper Name	Min. Gauge (In.)	Max. Width (In.)	Tensile Strength ksi (Min)	Elongation % in 2″ Min.
Annealed	.0014 .0028 or greater	25.5 25.5	20 25	10 20
As Rolled	.0014 .0028 or greater	25.5 25.5	50 50	1/2 1

ROLLED TEMPERS

Temper Name	Min. Gauge (In.)	Max. Width (In.)	Tensile Strength ksi	Elongation % in 2″ Min.
1/8 Hard	.0095	12	32/40	17
1/4 Hard	.0057	12	34/42	13
1/2 Hard	.003	12	37/50	5
3/4 Hard	.002	12	40/55	3

Figure 1.10b Mechanical property data for rolled annealed copper alloy 110. (Courtesy of Somers Thin Strip Division, Olin Brass, Olin Corporation, Waterbury, Conn.)

rolling mill called a Sendzimir mill (Figs. 1.14, 1.15). This mill will process the copper to a very even and consistent thickness while allowing a foil as thin as 1/2 oz/ft^2 (0.0007 in. thick) to be manufactured.

Rolled copper is extremely flexible and is very useful for applications requiring constant flexing for long periods of time in dynamic applications. High flexibility is obtained by producing copper foil with horizontal grain structure such as that shown in Figures 1.11-1.13. It is easy to see that the horizontal grain contrasts greatly with the vertical grain structure of electrodeposited copper. While the horizontal grain structure provides for excellent dynamics, its bonding surface is not as desirable as a material with a "tooth" or a vertical structure found in electrodeposited copper. The Somers

Flexible Circuits

PROPERTIES AS SUPPLIED
1 OUNCE

Temper	Tensile Strength (psi)	Elongation % in 2''
As Rolled	50,000 min	1/2 min.

PROPERTIES FOLLOWING TYPICAL ADHESIVE CURE CYCLE
FOR 1 OUNCE FOIL

Cure Temperature And Time (temp/min)	Nominal Tensile Strength (psi)	Nominal Elongation % in 2''	D_f [1]
350° F/15	21,000	12	52
350° F/45	19,800	11	56
350° F/180	20,600	11	58

[1] Parallel (MD) & Transverse (XMD) Average

Figure 1.10c Specific properties of LTA® rolled annealed copper alloy 110. (Courtesy of Somers Thin Strip Division, Olin Brass, Olin Corporation, Waterbury, Conn.)

Thin Strip Brass Group of the Olin Corporation manufactures a trademarked rolled copper foil called LTA copper. LTA is an acronym for low-temperature annealed. It is manufactured using an alloy 110 copper specially processed to provide a low annealing temperature. With this foil the copper can be handled and laminated by the flexible circuit manufacturer in a full hard condition, and yet will anneal to a very flexible material at normal adhesive curing temperatures and pressures. The advantage of this kind of product is that it reduces foil handling problems while still providing a highly flexible material. This results in higher yields and an attendant lower cost of the flexible circuit. Another copper treatment is usually provided to rolled copper to enhance the bonding characteristics of the product. One of the more popular treatments is rolled treatment K and is a proprietary process used by the Laminates Division of Oak Materials Group, Inc. This process involves the application of a thin deposit of copper onto the rolled copper surface using an electrolytic process. The final surface of the rolled copper with treatment K is similar to the surface of electrodeposited copper and enhances the bonding characteristics markedly.

The following pages include charts which should help the flexible circuit designer with the selection of the proper copper type for most applications (Fig. 1.10).

1.2.3 Adhesives

There are many kinds of adhesives or glues which are used to manufacture flexible circuit laminates. These include acrylics, epoxies, phenolic butrals, polyesters, and some Teflon combinations.

	WROUGHT-ANNEALED		ELECTRODEPOSITED FLEX GRADE	
Gauge (Weight)	.0014" (1 oz)	.0028" (2 oz)	.0014" (1 oz)	.0028" (2 oz)
Tensile Strength (psi) (NOMINAL)	25,000	34,100	40,000	44,000
Yield Strength (psi) 0.2% Offset (NOMINAL)	11,100	13,600	13,700	20,700
Elongation % in 2 Inches (NOMINAL)	17	35	12	18
Mullens Bulge Height	0.292"	0.363"	0.197"	0.267"
Electrical Resistivity (Weight: ohms-g/m²)	.152361		<.15940	NA

Figure 1.10d General comparison of rolled annealed copper with high elongation electrodeposited copper. (Courtesy of Somers Thin Strip Division, Olin Brass, Olin Corporation, Waterbury, Conn.)

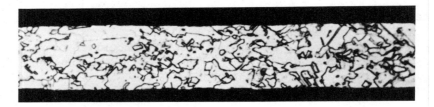

Figure 1.11 Micrograph of rolled annealed copper grain structure. (Courtesy of Somers Thin Strip Division, Olin Brass,
Olin Corporation, Waterbury, Conn.)

Figure 1.12 Scanning electron microscope photograph of
"HDO" treatment of rolled copper for higher bond characteristics. (Courtesy of Califoil, Inc., San Diego, Calif.)

Acrylics

Acrylic systems are generally used for flexible circuit applications with demanding temperature requirements such as those necessary for multiple-solder operations to avoid blistering or delamination, or when flexible circuits must withstand the repair or replacement of components several times. The most popular acrylic system used today is a product called Pyralux manufactured by DuPont. Table 1.6 shows the properties of this adhesive system when used with Kapton film.

Figure 1.13a Scanning electron microscope photograph of treatment K on rolled copper for higher bond characteristics. X3000. (Courtesy of Oak Materials Group, Inc., Oak Corporation, Hoosick Falls, New York.)

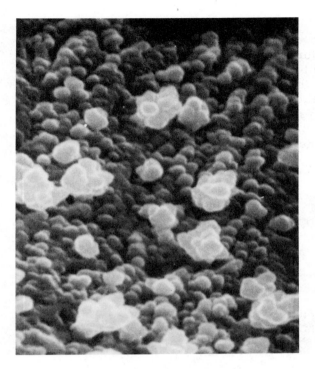

Figure 1.13b Scanning electron microscope photograph of treatment K on rolled copper for higher bond characteristics.
X90000. (Courtesy of Oak Materials Group, Inc., Oak Corporation, Hoosick Falls, New York.)

Epoxies

Epoxy systems are also widely used and are generally lower
in cost than acrylics. They have fairly high temperature resistance and can withstand the molten solder temperatures associated with dip-and-wave soldering. They are also excellent for use in applications which have extremely high environmental temperatures for long periods of time. Epoxy systems can withstand temperatures between 400 and 450°F for
many hours at a time without deterioration, whereas other systems such as acrylics will degrade over a long period of time
at these elevated temperatures. The bond strength of the
epoxy systems also improves somewhat after exposure to

soldering temperatures. Table 1.7 shows some of the characteristics of epoxy systems manufactured by Rogers Corporation.

Phenolic Butyrals

Phenolic butyrals are used in some systems which require high solvent resistance but must still give the flexible laminate a high degree of flexibility. Table 1.8 shows the characteristics of this type of adhesive system.

Polyesters

Polyester adhesive systems are used widely in some consumer applications where soldering temperature resistance is not a requirement and when it is necessary to manufacture laminate for flexible circuits using a polyester film as the dielectric. Typical applications of flexible circuits using polyester adhesives can be found in instant camera film interconnects and in

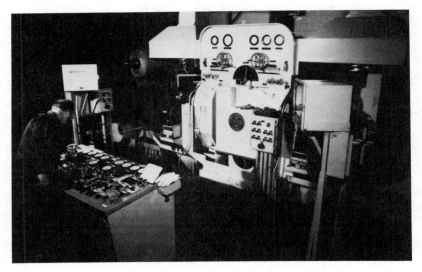

Figure 1.14 Sendzimir cold-rolling mill. (Courtesy of Somers Thin Strip Division, Olin Brass, Olin Corporation, Waterbury, Conn.)

Figure 1.15 Close-up of a Sendzimir cold-rolling mill. (Courtesy of Somers Thin Strip Division, Olin Brass, Olin Corporation, Waterbury, Conn.)

Table 1.6 Flexible Adhesive Bonding Films (Acrylic)

Material type: Unsupported acrylic adhesive
Material designation: 000M

Property to be tested and test method	Class 1	Class 2	Class 3
1. Peel strength, min., lb/in. width IPC-TM-650, method 2.4.9			
As received method A method B	6.0^a	8.0^a	8.0^a
After solder float method C method D	5.0^a	7.0^a	7.0^a
After temp. cycling method E method F	6.0^a	7.0^a	8.0^a
2. Low-temperature flexibility IPC-TM-650, method 2.6.18	N/A	N/A	Pass (5 cycles)
3. Flow, max. (squeeze out in mil) IPC-TM-650, method 2.3.17.1	5	5	5
4. Volatile content (max. %) IPC-TM-650, method 2.3.37	4.0	3.0	2.0
5. Flammability, min. % O_2 IPC-TM-650, method 2.3.8	15	18	20
6. Solder float method A IPC-TM-650, sec. 2.4.13 method B	N/A N/A	Pass N/A	N/A Pass
7. Chemical resistance percentage IPC-TM-650, sec. 2.3.2, method A	70	75	80

Table 1.6 (Continued)

Material type: Unsupported acrylic adhesive
Material designation: 000M

Property to be tested and test method	Class 1	Class 2	Class 3
8. Dielectric constant, max. (at 1 MHz) IPC-TM-650, method 2.5.5.3	N/A	4.0	4.0
9. Dissipation factor, max. (at 1 MHz) IPC-TM-650, m thod 2.5.5.3	N/A	0.05	0.05
10. Vol. resistivity, min. $M\Omega \cdot m$ IPC-TM-650, method 2.5.17	N/A	10^6	10^6
11. Surface resistance, min., $M\Omega$ IPC-TM-650, method 2.5.17	N/A	10^4	10^5
12. Dielectric strength, min. volts/mil ASTM-D-149	1000	1000	1000

13. Insulation resistance, min., MΩ IPC-TM-650, method 2.6.3.2 at ambient	10^2	10^3	10^4
14. Moisture and insulation resistance, min., MΩ IPC-TM-650, method 2.6.3.2	N/A	10^2	10^3
15. Moisture absorption, max. % IPC-TM-650, method 2.6.2	N/A	6.0	6.0
16. Fungus resistance IPC-TM-650, method 2.6.1	N/A	N/A	Nonnutrient

[a]Represents values for peel with bonding to treated side of copper. Values are halved when bonding to untreated copper surfaces.

N/A = Not applicable.

Source: From ANSI/IPC-T-50B. Used with permission of the Institute for Interconnections and Packaging Electronic Circuits, Evanston, Ill.

Table 1.7 Flexible Adhesive Bonding Films (Epoxy)

Material type: Unsupported epoxy adhesive
Material designation: 000L

Property to be tested and test method	Class 1	Class 2	Class 3
1. Peel strength, min., lb/in width			
IPC-TM-650, method 2.4.9			
As received			
method A method B	6.0^a	7.0^a	8.0^a
After solder float			
method C method D	5.0^a	6.0^a	7.0^a
After temp. cycling			
method E method F	6.0^a	6.0^a	8.0^a
2. Low-temperature flexibility			
IPC-TM-650, method 2.6.18	N/A	N/A	Pass (5 cycles)
3. Flow, max. (squeeze out in mil)			
IPC-TM-650, method 2.3.17.1	5	5	5
4. Volatile content (max. %)			
IPC-TM-650, method 2.3.37	4.0	3.0	2.0
5. Flammability, min. % O_2			
IPC-TM-650, method 2.3.8	15	18	20
6. Solder float			
IPC-TM-650, sec. 2.4.13 method A	N/A	Pass	N/A
method B	N/A	N/A	Pass
7. Chemical resistance percentage			
IPC-TM-650, sec. 2.3.2, method A	70	75	80

8. Dielectric constant, max. (at 1 MHz) IPC-TM-650, method 2.5.5.3	N/A	4.0	4.0
9. Dissipation factor, max. (at 1 MHz) IPC-TM-650, method 2.5.5.3	N/A	0.06	0.06
10. Vol. resistivity, min. $M\Omega \cdot m$ IPC-TM-650, method 2.5.17	N/A	10^6	10^6
11. Surface resistance, min., $M\Omega$ IPC-TM-650, method 2.5.17	N/A	10^3	10^4
12. Dielectric strength, min. volts/mil ASTM-D-149	500	500	500
13. Insulation resistance, min. $M\Omega$ IPC-TM-650, method 2.6.3.2 at ambient	10^2	10^3	10^4
14. Moisture and insulation resistance, min., $M\Omega$ IPC-TM-650, m thod 2.6.3.2	N/A	10^2	10^3
15. Moisture absorption, max. % IPC-TM-650, method 2.6.2	N/A	4	4
16. Fungus resistance IPC-TM-650, method 2.6.1	N/A	N/A	Nonnutrient

[a] Represents values for peel with bonding to treated side of copper. Values are halved when bonding to untreated copper surfaces.

N/A = Not applicable

Source: From ANSI/IPC-T-50B. Used with permission of the Institute for Interconnections and Packaging Electronic Circuits, Evanston, Ill.

Table 1.8 Flexible Adhesive-Coated Dielectric (Polyimide-Phenolic:Butyral)

Material type: Polyimide dielectric with phenolic butyral
Material designation: E1E L(1, 2, 3, and 5-mil thick dielectric)

Property to be tested and test method	Class 1		Class 2		Class 3	
	<0.002	≥0.002	<0.002	≥0.002	<0.002	≥0.002
1. Peel strength, min., lb/in. width IPC-TM-650, method 2.4.9						
As received						
method A method B	3	4	4	5	4	5
After solder float						
method C method D	3	4	4	5	4	5
After temp. cycling						
method E method F	3	4	4	5	4	5
2. Tensile strength, min., lb/in.^{2}						
<0.001	24,000		24,000		24,000	
IPC-TM-650, method 2.4.19						
<0.001	15,000		15,000		15,000	
3. Elongation, min. %						
IPC-TM-650, method 2.4.19	15		25		40	
4. Initiation tear strength, min. g						
IPC-TM-650, method 2.4.16	300		400		500	
5. Propagation tear strength, min. g IPC-TM-650, method 2.4.17.1						
<0.001 inch thick	5		5		5	
≥0.001 to <0.002	10		10		10	

⩾0.002 to <0.004	15	15	15
⩾0.004 inch thick	25	25	25
6. Low-temperature flexibility IPC-TM-650, method 2.6.18	Pass (5 cycles)	N/A	N/A
7. Density, min. g/cm^3 ASTM-D-772	1.3	1.3	N/A
8. Dimensional stability, max. percentage IPC-TM-650, section 2.2.4, method A	0.20	0.30	N/A
9. Flow (squeeze out in mil) Max. IPC-TM-650, section 2.2.4, method A	5	5	5
10. Volatile content (Max. %)	2	3	4
11. Flammability, min. % O$_2$ IPC-TM-650, method 2.3.8	20	18	15
12. Solder float IPC-TM-650, sec. 2.3.2 method A	N/A	Pass	N/A
method B	Pass	N/A	N/A
13. Chemical resistance percentage IPC-TM-650, sec. 2.3.2, method B	85	80	75
14. Dielectric constant, max 1 MHz IPC-TM-650, method 2.5.5.3	TBD	TBD	N/A
15. Dissipation factor, max. 1 MHz IPC-TM-650, sec. 2.5.5.3	TBD	TBD	N/A

Table 1.8 (*Continued*)

Material type: Polyimide dielectric with phenolic butyral
Material designation: E1E L(1, 2, 3, and 5-mil thick dielectric)

Property to be tested and test method	Class 1	Class 2	Class 3
16. Vol. resistivity, min., $M\Omega \cdot m$ IPC-TM-650, method 2.5.17	N/A	10^6	10^6
17. Surface resistance, min., $M\Omega$ IPC-TM-650, method 2.5.17	N/A	10^4	10^5
18. Dielectric strength, min. volts/mil ASTM-D-149	2000	2000	2000
19. Insulation resistance, min., $M\Omega$ IPC-TM-650, method 2.5.9	10^3	10^4	10^5
20. Moisture & insulation resistance, min., $M\Omega$, IPC-TM-650, method 2.6.3.2	N/A	10^4	10^5
21. Moisture absorption, max., % IPC-TM-650, method 2.6.2	N/A	N/A	TBD

Source: From ANSI/IPC-T-50B. Used with permission of the Institute for Interconnections and Packaging Electronic Circuits, Evanston, Ill.

Table 1.9 Flexible Adhesive Bonding Films (Polyester)

Material type: Unsupported polyester adhesive
Material designation: 000N

Property to be tested and test method	Class 1	Class 2	Class 3
1. Peel strength, min., lb/in. width IPC-TM-650, method 2.4.9			
As received			
method A method B	N/A	3.0	5.0
After solder float			
method C method D	N/A	N/A	N/A
After temp. cycling			
method E method F	3.0	3.0	5.0
2. Low-temperature flexibility IPC-TM-650, method 2.6.18	N/A	N/A	Pass (5 cycles)
3. Flow, max. (squeeze out in mils) IPC-TM-650, method 2.3.17.1	10	10	10
4. Volatile content (max. %) IPC-TM-650, method 2.3.37	2.0	1.5	1.5
5. Flammability, min. % O_2 IPC-TM-650, method 2.3.8	15	18	20
6. Solder float method A IPC-TM-650, sec. 2.4.13 method B	N/A N/A	N/A N/A	N/A N/A
7. Chemical resistance percentage IPC-TM-650, sec. 2.3.2, method A	70[a]	80[a]	90[a]

Table 1.9 (*continued*)

Material type: Unsupported polyester adhesive
Material designation: 000N

Property to be tested and test method	Class 1	Class 2	Class 3
8. Dielectric constant, max. (at 1 MHz) IPC-TM-650, method 2.5.5.3	N/A	4.6	4.6
9. Dissipation factor, max. (at 1 MHz) IPC-TM-650, method 2.5.5.3	N/A	0.13	0.13
10. Vol. resistivity, min. $M\Omega \cdot m$ IPC-TM-650, method 2.5.17	N/A	10^6	10^6
11. Surface resistance, min., $M\Omega$ IPC-TM-650, method 2.5.17	N/A	10^4	10^5
12. Dielectric strength, min. volts/mil ASTM-D-149	1000	1000	1000
13. Insulation resistance, min., $M\Omega$ IPC-TM-650, method 2.6.3.2 at ambient	10^3	10^4	10^5
14. Moisture and insulation resistance, min., $M\Omega$ IPC-TM-650, method 2.6.3.2	N/A	10^3	10^4
15. Moisture absorption, max. % IPC-TM-650, method 2.6.2	N/A	2.0	2.0
16. Fungus resistance IPC-TM-650, method 2.6.1	N/A	N/A	Nonnutrient

[a]Except chlorinated solvents and ketones.
N/A = Not applicable.
Source: From ANSI/IPC-T-50B. Used with permission of the Institute for Interconnections and Packaging Electronic Circuits, Evanston, Ill.

automobiles as the instrument cluster connection system. Table 1.9 shows characteristics of polyester adhesive systems.

Other Adhesives

A typical fusion adhesive system is one wherein FEP or a combination of FEP/TFE is used to bond a substrate such as Kapton to copper. This system has extremely high temperature characteristics, however, temperatures of over 500°F are necessary to obtain a good bond between the films and foils. This kind of system is used primarily in military applications and does not represent a high percentage of the flexible circuit applications available. These materials are also difficult to work with since cover films using this system must also be laminated at extremely high temperatures allowing the conductors which have been previously etched in the base materials to "swim," due to the remelting of the base laminate adhesive. Other adhesive systems get around this problem by cross linking at the curing temperatures, and subsequently, requiring much higher temperatures to remelt. Designers should discuss various adhesive systems with their flexible circuit vendor to optimize the system to the specific flexible circuit application they have in mind.

2
Production Processes

Carl A. Edstrom, Jr.
Flexible Circuits Business Unit, Sheldahl, Inc., Northfield, Minnesota

2.1 INTRODUCTION

The production process utilized in the manufacture of flexible printed circuits includes many steps and a variety of materials and chemicals dependent on design specifications. All of the stresses inherent in the manufacturing process come to bear during the production phase. Today's customers demand zero or near-zero defects. As a result, controlled and predictable processes, coupled with 100% testing of production runs are required. High quality, cost containment, and de-escalation add to today's manufacturing challenge.

2.2 VERIFICATION OF RAW MATERIALS

The verification process assures that the product or the component of the product or the tool that is used on the product is at or beyond the specification required. This will help insure that the finished circuit not only works, but also meets all test and performance specifications. Trouble can arise in two areas in the raw materials verification process: (1) the specification which defines the raw material characteristics and the (2) processability of a specific batch or lot of raw material. Ideally, these two are combined in the material specification. Processibility tests are only required when compatibility with another material is in question. For example, copper surface topography as it relates to a unique lot of adhesive may require processibility testing.

There are five basic classes of materials which are used in the production of flexible printed circuits: foils, dielectrics (film/coatable liquids), plating materials, adhesives, and hardware and components.

2.2.1 Foils

Electrodeposited and rolled copper foils are normally used in flexible circuitry. These foils should be checked for thickness, purity, tensile, elongation, and dimensional ductility.

The Mullen bulge test measures three-dimensional ductility. A piece of copper foil is placed over a rubber diaphragm; the diaphragm is then loaded hydraulically, inflating much as a balloon does. The height of the diaphragm at which the copper bursts (the bulge) is then measured. This test may or may not be used, depending on the group producing the product.

This is an empirical test combining tensile and elongation in a three-dimensional way. It is an excellent test for determining the bendability of copper. Since flexibles must bend, naturally one must know how tolerant the design is to bending. The characteristics of bendability can vary from batch to batch, and whether electrodeposited or rolled foil is used. An increasingly common conductive material is silver-loaded epoxy ink. Common verification tests for this ink include viscosity, mixture analysis, particle size, flexibility after cure, and line conductivity when printed.

2.2.2 Dielectrics (Film/Coatable Liquids)

Some basic properties of dielectrics (polyester films, polyimide film, aramid fiber paper, and composite films) are as follows:

Polyester Film

   1.   Excellent electrical properties—insulation resistance

      a. 7500 V/mil for one-mil film
      b. Surface resistivity $10^{16}\Omega$
2. Low moisture absorption
      a. Less than 0.8%
3. Good thermal stability, capable of operation up to 150°C
      a. Melting point is 250°C (480°F)
4. Good dimensional stability
      a. 0.5% at 150°C
      b. 0.2% when prestabilized
5. Chemical resistance to common solvents
6. Cheapest of the commonly used materials
      a. 2 cents/ft^2 per mil*

Polyimide Film

1. Highest temperature capabilities
      a. Zero strength temperature = 815°C
      b. Melting point, none
      c. Usable up to 400°C
2. Excellent electrical properties—insulation resistance
      a. 7000 V/mil for one mil
      b. Surface resistivity $10^{16}\Omega$
3. Excellent chemical resistance
      a. Resistant to most common solvents
      b. Degraded by sodium hydroxide
4. Relatively high moisture absorption
      a. 2.9% maximum
5. Good dimensional stability
      a. 0.3% shrinkage at 250°C
6. Most expensive of the commonly used materials
      a. 33 cents/ft^2 per mil.*

Aramid Fiber Paper

1. High-temperature material, capable of continuous operation at 220°C
      a. Has a UL rating

---

*These numbers may vary slightly depending on whether they are obtained from the material supplier, the circuit salesman, or the circuit manufacturer.

  2. Good electrical properties—insulation resistance
      a. 530 V/mil for 2-mil paper
  3. High moisture absorption
      a. 2.9% at 50% relative humidity (RH), 2-mil paper
      b. 7.7% at 95% RH, 3-mil paper
  4. Dimensions dependent on moisture content
      a. 4-5 mil/inch expansion from dry condition to 50%
          RH, 3-mil paper
  5. Intermediate cost is between polyimide and polyester
      a. Approximately 8 cents/ft^2, 2-mil paper*

Composite Dielectrics

Composite dielectrics are similar to polyimide film but have
better dimensional stability and lower moisture absorption
characteristics. These dielectrics are usually 0.004 and 0.007
in. thick with flexibility characteristics similar to thin G-10
epoxy glass material.

2.2.3 Plating Materials

Verification of plating materials generally involves testing for
the composition or purity of the metal to be plated, concentration of plating solutions, and purity of the anode. Copper
anodes, for example, are purchased with trace concentrations
of phosphorus included, but must be free of all other impurities. Plating materials are usually received from the manufacturer with certification to a specification. In the production process, the materials should be periodically audited to
insure compliance with design specifications.

2.2.4 Adhesives

Most adhesives are polymers of polyester, epoxy, phenolic, or
acrylic. Therefore, the verification process involves determining composition, checking melting points and viscosity,

---

*These numbers may vary slightly depending on whether
they are obtained from the material supplier, the circuit
salesman, or the circuit manufacturer.

pretesting for laminating temperatures and parameters, and specification for temperature resistance in the soldering environment.

2.2.5 Hardware and Components

Mechanical devices such as eyelets, connectors, and clips and electronic components such as resistors, capacitors, diodes, and fuses are commonly attached to flexible circuits. Normal tests to characterize these materials include tests for hardness, spring rate, plating type, thickness, and retention, along with the operating characteristics of the operating components. Circuits with mechanical fasteners attached should be qualification-tested under tougher than normal environmental conditions (i.e., accelerated aging conditions) to give adequate assurance of reliable functional capability.

2.3 VERIFICATION OF PROCESS MATERIALS

The commonly used process materials in the production of flexible circuits are etchants, solvents and cleaning solutions, resist stripping and developing materials, and inks. These are the materials that are used to produce circuits, but do not become an integral part of them. In the production of flexible circuits, these process materials may be significant cost items.

2.3.1 Etchants

Concentration is the key to how fast the etchant acts. The proper temperature and time parameters to achieve the correct etch rate are indicated in the etchant's concentration. Common etchants used for copper are ammoniated alkaline etchants, cupric chloride, ammonium or hydrogen persulfate, peroxy sulfurics, and ferric chloride.

2.3.2 Solvents and Cleaning Solutions

Production solvents are tested for impurities, trace metals, and concentrations to assure appropriate activity during the production process. These materials must be carefully

matched and controlled to suit the specific operation or desired function.

2.3.3 Resists

Etching and plating resists are primarily checked for viscosity at production temperatures. Viscosity under shear conditions (a measure of thixotropic properties) may also be a critical test of resists. A resist's flow properties define its ability to effectively screen print an appropriate pattern. Too
much or too little flow can create equally serious problems.
Cured film flexibility, cure times and temperatures, adhesion,
chemical resistance, stripping characteristics, and cured film
thickness and porosity are othe key considerations when
verifying resists.

2.4 VERIFICATION OF TOOLS AND FIXTURES

There are three basic types of tooling utilized in the manufacture of flexible circuits: artwork, steel tools (soft and hard),
and assembly and test fixtures. It does little good to check
the purity and quality of the raw materials, verify the characteristics and properties of the process materials, and com-
bine them to form products using inappropriate or improperly
set up tooling and fixtures.

2.4.1 Artwork (Image and Coverlay)

Artwork for flexible circuitry applications must be as good
and sharp as possible and constructed and nested very accurately. If an application has a total manufacturing tolerance of 0.015 in. for the image to the pierce pattern, it cannot tolerate the loss of 0.005 in. of that tolerance in the construction of the artwork package. In addition, for low-cost
manufacture of circuitry, the artwork features should be op-
timized over the available area to allow as much manufacturing
tolerance as possible. Artwork is most often provided by the
customer; it should be checked often and its status and quality reported back to the customer as necessary. If a screened
cover coat or solder stop (or nomenclature) is sued, artwork
must be prepared for these steps also.

For production jobs where the potential for repeat orders or significant volume is high, master artwork is often committed to glass for security and stability.

Artwork is usually checked dimensionally, from center-to-center dimensions of holes, from center-to-center dimensions of conductors, conductor width, and feature dimensions relative to the entire pattern.

Computer-aided dimensional checking equipment is the industry standard today. Its use guarantees very accurate artwork patterns which can be coordinated to the dimensional features of the steel tooling for the product. (Fig. 2.1.)

2.4.2 Steel Tools (Soft and Hard)

The steel tools used in the manufacture of flex circuits consist of registration hole-piercing dies (index), coverlay-piercing dies, compound pierce-and-blank dies, progressive dies, or steel rule dies, as applicable for a specific job.

Pierce-and-blank dies consist of a pattern of punches and an outline blank. This type of die is contained within a single die cavity and defines all internal holes and features and the outline all in one operation. It produces all of a product's dimensional features in very accurate relationship to each other.

A variation of this die is the progressive die. This die has two or more piercing and flash or cutting stations that operate in sequence rather than in unison. Thus, the material or the product travels through each die station in sequence. For example, a circuit coming into a progressive die might first go through the piercing station where all internal holes and other features are cut, and then into the station where the outline would be cut. Progressive dies are used when product features are so close together that they cannot be built in the same die. They are also used for high-volume applications (possibly automated) to minimize the necessity to handle the piece part and remove it from the web.

Another type of tool, a soft steel tool, is referred to as a rule die. The product is laid on top of the cutting surface and a pressure pad is actuated and the product cut, much like a cookie cutter. Rule dies cannot put very small (component mounting) holes in circuits. Consequently, they must

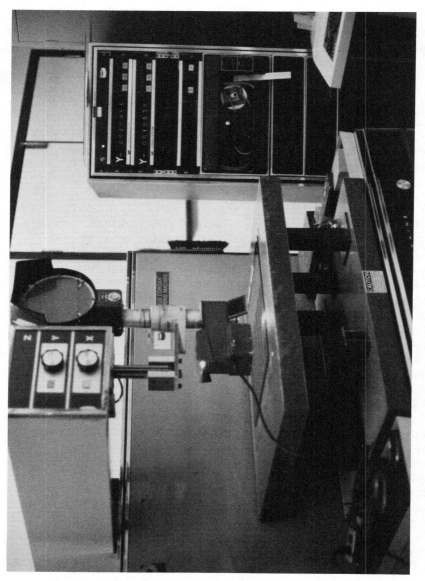

Figure 2.1 The Cordax machine is used in the verification of artwork from the customer; it electronically measures the image and coverlay specifications of the product to be produced in very precise detail. (Courtesy of Sheldahl, Inc., Northfield, Minn.)

usually be used in conjunction with numerically controlled (NC) (or scope) drilling of these small holes.

2.4.3 Assembly and Test Fixtures

Assembly fixtures which need verification are the holding devices, jigs, and bending devices required to help form circuits or to position film coverlays over the prepared circuitry patterns. This equipment is checked to insure production of the appropriate alignments required by the manufacturing process. During the production process, fixtures are used for creasing, bending, or twisting the flexible circuits; fixtures are also used for alignment of the circuit to a rigidizing board while the flexible circuit is bonded to specific areas of the rigid board.

The drill or routing tapes also need verification. Much of the equipment is numerically controlled and computer-driven. They use generated tapes that drive the machines. Tapes may also be generated which drive routing heads on similar types of machines.

2.5 LAMINATION OF METAL-CLAD DIELECTRIC

Following the verification of raw materials, artwork, gages, tools, and fixtures, the manufacturing process actually begins. The first step is the laminating and coating process to develop the materials that are used in the fabrication of flexible circuitry. The laminate is a metal-clad dielectric usually made of copper, an adhesive layer, and a dielectric film. The coverlay is adhesive-coated dielectric film. (Fig. 2.2.)

2.5.1 Single- and Double-Sided Laminations

The laminate is produced through a process that utilizes a liquid adhesive to coat the film or foil. There are a variety of ways to coat the adhesive on the copper foil: reverse roll, pressure bar, gravure, or curtain coating. The adhesive-coated material is then run through a driving tunnel, which removes a predetermined amount of solvent from the adhesive and dries the remaining solvent to a tacky state. Next, the material is run through combining rolls, which use heat and

Figure 2.2 This roll-to-roll laminator adheres materials together that will be used in the production of the flexible circuitry. (Courtesy of Sheldahl, Inc., Northfield, Minn.)

pressure to bond the layers together. The end product is a cross-section of film, adhesive, and foil, typically 0.005 inches thick. The process just discussed would result in a laminate for a product which is only one layer of conductive copper circuitry. If two layers were necessary, the process would be repeated once again.

2.6 ADHESIVE COATING OF DIELECTRIC FILM

Adhesive coating of dielectric films is another process utilized in the making of flexible circuit materials. Instead of coating copper in this process, the dielectric film is coated with adhesive and then wound on a large roll. The resulting product is adhesive-coated film which is used as the top insulating layer for the circuit. The coverlay features are then defined and bonded onto the circuit after the circuit has been fabricated.

2.7 CASTING OF UNSUPPORTED ADHESIVES

A third process which is really a variation of adhesive coating is the casting of adhesive on a nonsticking release film to produce unsupported bonding plys, or the casting of liquids directly onto the foil as the dielectric layer of the laminate.

2.7.1 Bonding Ply

A variation of coating adhesives onto film is applying adhesives to a material to which it will not stick when it is dry. The end product of this process is an adhesive film which is used to bond discrete flexible circuits together during final assembly. There is no dielectric on it. This kind of bonding is used when the flexible circuit is built separate from the assembly and needs to be adhered to a carrier.

2.8 SLITTING AND SHEETING

The final process in the preparation of the raw materials used in the fabrication of flexible circuitry is a slitting and/or sheeting operation. Typically, laminating and adhesive coating are done at mill width in mill-length runs. Film and foil are purchased in 25-50-in. widths and then slit to the necessary width for the application being done or cut into sheets if the sheet process is to be used. In addition, some adhesive types of material require platen-press curing in sheet form to develop full bonds and chemical resistance.

2.9 INDEXING

The manufacturing of the actual circuits begins with the indexing process. This is used to define manufacturing registration or manufacturing reference holes for all subsequent processes.

2.9.1 Step-and-Repeat Manufacturing of Alignment Holes

Indexing is necessary to align the printing, the coverlays, and the blanking when a step-and-repeat pattern is used. This is accomplished by the drilling or punching of a series of holes which are used for alignment of all future operations in the manufacturing process (Fig. 2.3).

2.9.2 Continuous Perforating of Manufacturing Alignment Holes

Continuous perforating of manufacturing alignment holes is similar to indexing, except that the process is continuous, as in the edges of prepunched computer paper or a roll of photographic film. This process generally is used for very-high-volume applications where significant automation equipment is available and can be utilized.

2.9.3 Positioning of Optical Alignment Targets

Target marks are printed on the copper side of the laminate as a part of the imaging process. These are used to assure proper alignment as manufacturing proceeds.

Figure 2.3 The Martin indexing press punches a series of alignment holes in the laminate. (Courtesy of Sheldahl, Inc., Northfield, Minn.)

2.9.4 Producing Via Holes for Two-Sided Plated-Through Hole (PTH) Circuitry

In an indexing process, there are two additional steps which might take place. If a plated-through hole (PTH) circuit is called for in the design, via holes are produced in the material, either through a punching process using a step-and-repeat process and a piercing die or, with a drilling process utilizing NC drill. Plated-through hole circuitry basically refers to the technology used to interconnect two separate planes or levels of circuitry. A process is utilized that chemically deposits a very thin layer of paladium across the nonconductive or insulating surface. Copper is then electrolessly

Figure 2.4 The punch press is used in the indexing process to put small holes in the material for plated-through hole circuits. (Courtesy of Sheldahl, Inc., Northfield, Minn.)

plated onto this surface to make the electrical connection between one plane and the other. In later processes when current crosses these two planes as they are put into a plating
bath, additional thicknesses of electrolytically plated copper
begin to build up. This improves the connection between the
top and bottom layers of the circuitry. Copperplating across
the inside of the small hole between the two layers of circuitry
will result from this process. The copperplating in the hole
is referred to as the barrel of the hole, the copper barrel.
To ensure that the hole is actually there, and in the proper
place, the holes are punched or drilled. This is in addition
to the mechanical registration holes that are punched or
drilled (Fig. 2.4.).

2.9.5 Punching Access Holes for One-Sided Back-Bared Pad Circuitry

In single-sided circuits, where copper is exposed on both
sides, another adaptation of indexing is used. This is called
back-bared circuitry, that is circuitry which is bared on the
back side as well as the top side. Adhesive-coated film which
has been punched or drilled with a hole pattern and then
bonded to copper is utilized here. The resultant circuit
would include a conductive pattern that was insulated top and
bottom except for those places in either film where holes were
cut. If exposure of some of the circuits on the back side are
called for in the design, holes must be punched out before
the copper is laminated to it. The resultant product is a
single-sided circuit to which components can be assembled on
either side. When a single-sided circuit is called for with access on both sides, a registration and access hole would be
punched in a film on the outside edge; copper would be bonded to the center of the punched-film laminate. The external
registration holes would then be utilized to align the process
when the coverlay is to be put on. An alternative process
that can be used to accomplish the same end result, but is
cost effective only on low volumes, is "sciving" away or mechanically removing the insulating film and adhesive in specific areas on one or both sides of the circuit after lamination.

2.10 IMAGING

Image placement graphically defines the conductive pattern of the circuit. There are two methods to achieve this process: screen printing and photoexposing.

Screen printing involves either a positive or a negative treatment on the copper. The treatment is dependent on whether the design calls for the copper pattern to be protected through an etching process or for the copper to be exposed to a plating process. By screening a positive, the desired copper pattern itself is protected, then put into etching, resulting in a discrete conductive pattern. When screening a negative, the offal or copper, which is not the desired circuitry pattern, is protected; this eventually will be etched away. The exposed circuitry copper pattern is then plated with 60/40 solder or gold in a conductive pattern. The negative pattern resist is then stripped off and the product is etched. The plating acts as an etch resist.

2.10.1 Screen Printing (Step-and-Repeat)

Screen printing can be done in a variety of ways, the most common of which is the step-and-repeat process. In this process, a discrete pattern is screened on the laminate. The screen printer is a machine that utilizes a frame covered with a fine mesh on which the image of the conductor pattern is photographically imposed. The mesh is held in such a way that it can be raised or lowered onto the copper-clad laminate. These machines also feature a squeegee that travels back and forth over the screen mesh, creating a pumping action through the screen. This produces a controllable thickness of resist to be deposited in the desired pattern on the copper. Normally, a drying or curing apparatus is used with the screen printer. This apparatus may be a drying tunnel, forced air, gas fired, electric, or steam, or in the case of processing sheets of material that can be racked up and pushed into a walk-in type oven.

Mose screen-printable materials used today are solven-based and require a process used simply to drive out the solvent; or, in the case of some of the epoxy-based materials, a process is necessary to cure the epoxy.

Printing machines can be manual, semiautomatic, or automatic, either sheet stock or material webs (roll-to-roll). There are a variety of single-color printers and multicolor printers. Basically, each of these machines utilizes standard screen-printing technology. The use of ultraviolet (UV) curing materials with appropriate UV-resist-curing units is a newer technology now gaining significant use (Fig. 2.5).

Photoexposed Imaging

Photoexposing may also be used to develop the imaging process. This basically consists of laminating a photosensitive emulsion onto the surface of the copper. This photosensitive emulsion may be either a dry film that would be laminated or a liquid which would be flood-coated. The material is then put through an exposing and developing process, much like that used in photography. Dry-film photosensitive material is most commonly used because of its ease in handling. The laminated material is then inserted into an exposer which, when coupled with a black and clear pattern of artwork, exposes those portions of the photoresist that are not covered by the artwork-emulsion pattern. This process cures that portion of the photoresist exposed to the light, and an image is created.

The differences between screen printing and photoexposing are related to the cost and speed of each process and to the control and precision of image definition. Photoexposing is much more expensive due to the cost of the photo emulsion and all subsequent operations: the exposing, the developing, and the stripping. However, it is quicker and easier to set up a run with exposing than with screening. The quantity to be run at a time thus becomes a significant factor as there is a tradeoff between set-up time and cost versus actual cost differences of the imaging operations themselves (Fig. 2.6).

Print/Etch/Plate Resist, UV Cover coat, Graphite, Nomenclature, Etc.

When working with imaging, cost as well as speed are considerations. In low-to-medium volume jobs, turnaround time is the key; photo processing does not require as much forethought and planning as screening. Another major

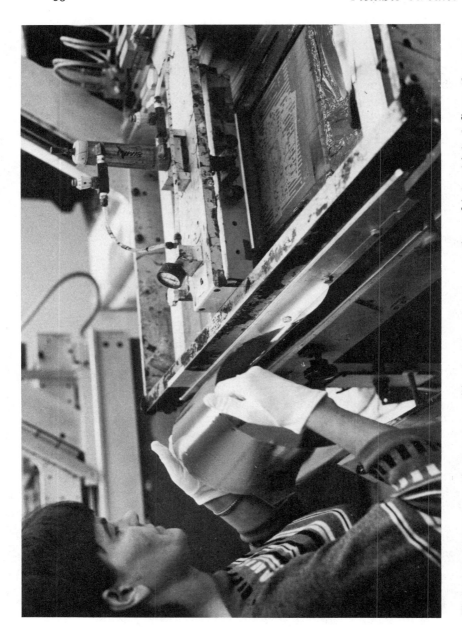

Figure 2.5 Screen printers are utilized to lay down a coating of ink which defines the conductive pattern of the circuit. (Courtesy of Sheldahl, Inc., Northfield, Minn.)

Figure 2.6 Imaging can also be produced on the laminate through the use of a photo exposer. Photo exposers are operated in clean rooms, and use a process similar to that used in photography. (Courtesy of Sheldahl, Inc., Northfield, Minn.)

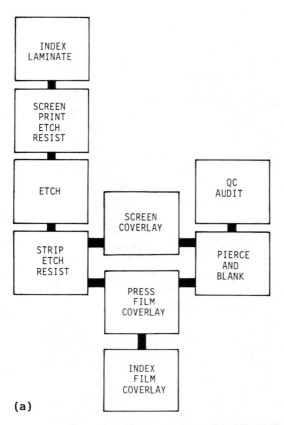

(a)

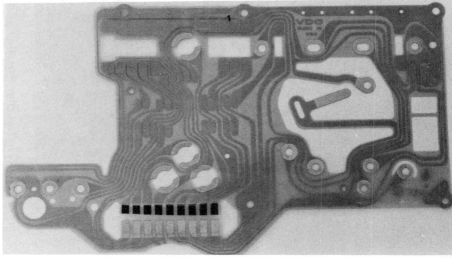

(b)

consideration is clarity or definition of the pattern. The photo-resist method allows very accurately defined lines and spaces relative to each other. It is more accurate in terms of feature definition and ability to define a top pattern in close registration to a bottom pattern on a plated flexible circuit. The dry film used in photo processing has line width and space definition capability down to 0.004 in.; screen printing has a capability down to 0.010 in. on conductors and spaces. Thus, when working with dense circuitry, image definition capability becomes a major consideration and when applicable, photo-processing techniques should be utilized (Fig. 2.7).

Beyond defining the image, there are a variety of other features that are added in a screen-printing operation. Cover coats are normally used in those applications when the circuit is going to be wave-soldered and solder is not desired on the entire board. Basically, cover coats are protective insulating films normally consisting of ultraviolet screen-printable materials. Nomenclature may also be printed on the circuit if required.

In some applications of screen printing, graphite is used in conjunction with the copper pad. The graphite is milled into minute particles, mixed with a carrier, usually an epoxy, and used in the printing process in the same manner as a resist. The graphite protects the copper pads from oxidizing in applications where plated surfaces are not desired.

2.10.2 Die Stamping

There are several other ways to define images. Die stamping is used primarily for single-sided circuitry and for power circuitry. The process involves developing a die that has cutting edges in the form of the conductive pattern called for in the design; the copper is then embossed or cut onto the dielectric, the excess copper peeled away, and the pattern then heat-laminated to the base dielectric.

Figure 2.7 Print-and-etch with coverlay. (a) Flowchart, (b) model. (Courtesy of Sheldahl, Inc., Northfield, Minn.)

2.10.3 Additive Circuitry

Another method to define images is to use chemical additive processes whereby the copper conductors are added electro-chemically to a piece of dielectric film. Basically this process utilizes a plastic film that is photosensitized so that selected portions of the film can be plated. This chemical additive process is still fairly new, but is being used increasingly by the industry. This process is a desirable alternative for those products calling for only very thin layers of copper.

2.11 ETCHING

Etching is the next step in the production process of flexible circuitry. The machines used are called etchers. They vary from single-chamber machines to etchers with numerous chambers and spray systems.

2.11.1 Types of Machines

The basic etching machine is a device that conveys either sheets of material or rolls under a series of nozzles that spray etchant on the sheet or roll, or, if an immersion etcher is used, the material is immersed in a liquid bath. The etchant is a chemical that attacks and dissolves the exposed metal.

After the circuit has gone through the etcher, the metal protected by the resist remains and any excess metal has been etched away. Spray etchers and closed-loop regeneration-type etchers are most often used to allow for control and monitoring of temperature and concentration. These two factors define how fast etching can be done.

In a multiple-chamber etcher, each part of the material goes through several distinct areas that perform part of the etching process. In a five-chamber etcher, the copper in the spacing area will break at the beginning of the last chamber and finally etch down through to the base. The etching is completed in the final chamber.

Immersion etchers are used for very fine line, close-tolerance etching. This process involves immersion in the liquid etchant. Immersion etchers are much slower, more precise, and more controlled than spray etchers.

There are a variety of etchants available, most of them geared to use with copper. The different etchants have varying etch rates and factors* and vary in compatibility with a range of metals. They also differ in terms of safety and pollution control (Fig. 2.8).

2.11.2 Resist Stripping

Following the etching process, the material is then stripped, a chemical process that removes the etching resist. Stripping is done with a spray bar or a dip in a bath. Mechanical scrubbing may also be used, but is not recommended much today because it tends to abrade the copper surface and leave stress-concentrating scratches in the copper conductors.

2.11.3 Antitarnish Application

Mechanical scrubbing involves a roller device, usually in water, or an antitarnish chemical that mechanically breaks the surface of the circuit, making it shiny; but, it also changes the structure of the material. When utilizing copper, the mechanical scrubbing technique is used as a last resort. But on plated solder, the mechanical abrasion tends to smooth out or flatten the solder and results in a longer shelf life; it also allows for more wettable solder. When used on solder-plated products after etching, care must be taken to prevent distrubing the overhanging solder edges (result of undercut in etching). If these break loose, solder slivers result.

---

*Etch factor or undercut factor is a ratio of how fast an etchant penetrates downward compared to how fast it penetrates sideways (under the resist).

Figure 2.8 Excess metal on the laminate is eliminated in the etching process. This is a roll-to-roll multichamber spray etcher. (Courtesy of Sheldahl, Inc., Northfield, Minn.)

2.12 PLATING

2.12.1 Batch and Continuous Roll-to-Roll

Plating is another step in the production process. There are
two basic plating processes, batch plating and continuous roll-
to-roll plating. Batch plating is the plating of sheets of material or discrete parts of material; this process can be manual or automatic (Fig. 2.9).

2.12.2 Automatic Batch Plating

Most common in the automatic area is the automatic rack line.
This process consists of a series of tanks of chemicals that
clean, sensitize, plate, and rinse the material. This is done
to plate a metal or an alloy onto the copper to enhance solderability or minimize tarnishing. In the case of gold, this

Figure 2.9 Continuous roll-to-roll solder platers electrochem-
ically plate a metal or alloy to the copper pattern. (Courtesy
of Sheldahl, Inc., Northfield, Minn.)

process is used to develop a good electrical surface for a switching or soldering application.

The continuous roll-to-roll plating process is chemically the same as the batch-line process, but has a material transporter built into it so th t rolls of material can be rolled out of one tank and into the next.

2.12.3 Types of Plating

Gold, Silver, Platinum, and Copper

There are different types of metals that are used in the plating process. Gold, silver, and platinum are precious metals that are often used. When using immersion gold in the plating process, the copper material is put into a gold bath that ionically exchanges surface copper for a very thin layer of gold. When all of the copper is covered by gold, the plating process stops, and the result is about 8-10 millionths of an inch of gold is deposited on the material. There is no nickel barrier underneath. In electroplated gold, either soft or hard gold may be utilized. The two types of gold differ in the kind of bath used. Hard gold is run through a cyanide bath. When cyanide gold is deposited, an electrical current is used to do the actual plating. Most often a nickel flash is plated between the copper and the gold to minimize gold migration into the copper. Plating can also be done with silver. Tin plate is sometimes used instead of solder. Tin is solderable but tends to oxidize quickly (Fig. 2.10).

Tin and Lead Solder Alloys

A variety of tin/lead solder alloys are commonly used as plating surfaces on circuitry. Eutectic solder or 63/37 solder (the lowest melt point tin/lead solder), is called for in many designs, however, 60/40 solder is often used and is compatible with most applications. When two soldering operations are very close together, different melt-point solders must be used so that the integrity of both solder joints can be maintained (Fig. 2.11).

Figure 2.10 Gold, an excellent conductor, is sometimes used in the plating process. This is a goldplate line in which the copper conductors are plated with a thin layer of nickel and gold. (Courtesy of Sheldahl, Inc., Northfield, Minn.)

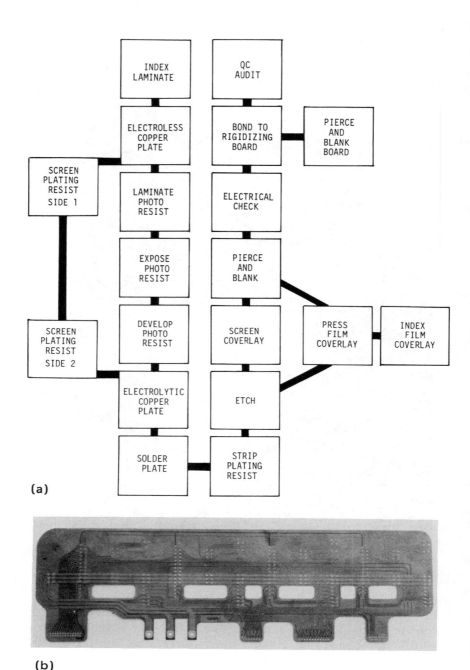

(a)

(b)

Figure 2.11 Two-sided pattern-plated-through hole circuit with coverlay. (a) Flowchart, (b) model. (Courtesy of Sheldahl, Inc., Northfield, Minn.)

Electroless and Electrolytic Copperplating

Besides the cover platings already discussed, electroless and electrolytic copperplating are used for two-sided plated-through hole circuitry. The electroless copperplate is a chemical exchange of copper for paladium. The paladium bearing seeded material is dipped into a copper bath which is warm but has no current. The chemistry of the bath is such that the copper in solution exchanges with the paladium in a sensitizing layer to give a seed coat, or a few millionths of an inch of copper in the through hole. This can then be electrolytically plated up. Plating of copper can either be done through panel plating or pattern plating. This refers back to the imaging process. The process that would be used for plated-through holes is one where the negative of the conductor pattern is put on the copper, and then the conductor pattern itself is plated up with copper. In the case of a two-sided plated-through hole circuit, the conductor pattern is plated up as the through holes are plated through. This is a pattern plating technique where copper is added only to the conductor areas of the circuit. A variation of this technique is panel plating, where the entire surface of the laminate is plated first. In this case, a positive artwork pattern for image definition is used and etched immediately after imaging (Fig. 2.12).

Pattern vs. Panel Plating

In pattern plating, two-sided plated-through-hole circuitry, the actual process involves first putting on the electroless copper, usually a few millionths of an inch thick. A very thin layer of electrolytic copper is then plated over the electroless copper to hold it in place. Next, a negative image is added using a screen printer or photo processor. The material would be plated-up electrolytically to achieve the barrel thickness desired. Solder plating would then be added to the plated copper and this acts as the etch resist. The imaging resist is now stripped and the circuit etched.

In panel plating, electroless copper would be put on first, then plated with electrolytic copper to the desired thickness over the entire laminate surface and through the holes. Next, the materials are put through the imaging process in which a

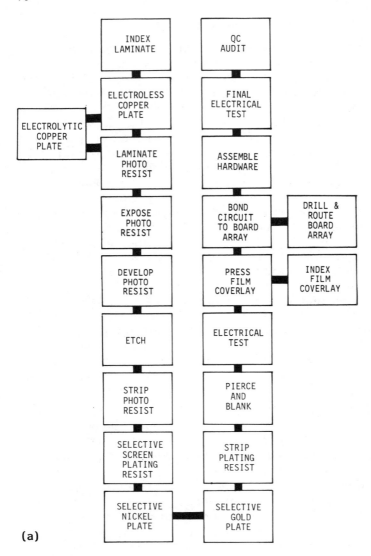

Figure 2.12 Two-sided panel-plated through-hole circuit with selective nickel/gold plating, film coverlay, rigid board array, and hardware. (a) Flowchart, (b) model. (Courtesy of Sheldahl, Inc., Northfield, Minn.)

positive of the conductor pattern is used to protect the copper conductors as well as the through holes. The waste is
then etched away. The resulting circuit is plated through,
but has no solder on it.

2.12.4 Selective Tab Plating

Tab plating is a variety of plating that is usually confined to
precious metals. It is a process that was historically developed to plate the output fingers on circuit boards with gold.
In the flexible circuit area, selective tab plating is done with
vertical strip platers. Vertical strip platers take a web of
material, tip the material on its edge, and dip plate the end
of it in a vertical format.

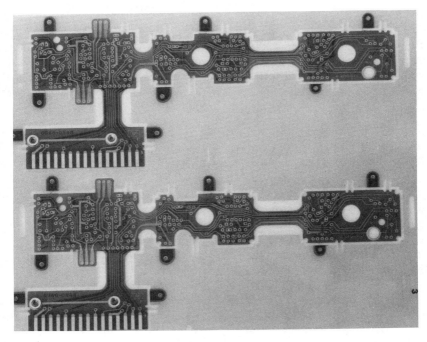

(b)

Figure 2.12 (Continued)

2.12.5 Additive Processing

Fully additive plating has been used very little for flexible circuits. It begins with bare dielectric material that is selectively sensitized. Electroless plating can then be utilized to build-up only the desired circuitry pattern. This is a slow and relatively costly process. It will become more attractive as thinner conductors gain wider acceptance.

2.12.6 Roll Tinning

Roll tinning applies molten solder to a circuit pattern using rollers partially submerged in solder. The rollers transfer the solder to the circuit surface in thicknesses from 0.0001 to 0.0025 inches.

2.12.7 Solder Brightening

Solder brightening is a chemical operation that removes stains from solder materials. It is a cosmetic operation which makes soldered surfaces brighter and increases solderability and shelf life.

2.12.8 Solder Scrubbing

Solder scrubbing is a mechanical brushing operation that smoothes the surface of the solderplated copper.

2.12.9 Solder Leveling

There is solder-leveling equipment that will take either molten solder which has been applied to a circuit or plated solder and melt and level it. These processes are supportive to the function of getting the solder onto the circuit.

2.13 COVERLAYING PROCESSES

The coverlay, found on the top side of the circuit, protects the conductor pattern from external environments. This insulating material can either be a discrete film or a screen-printable material (a conformal coating). In the latter, they are predominantly ultraviolet curable materials.

When utilizing film coverlays, the pattern of the coverlay, or its access holes, are either drilled into a sheet of adhesive-coated film, or they are punched with a pierce die. The drilling or punching defines the size of the hole in the coverlay film that will expose the conductive pad. Thos adhesive-coated films which are punched or drilled are then mechanically aligned to the circuit and bonded or laminated to the circuit with heat and pressure in a platen press.

2.13.1 Coverlay Films: Encapsulation

Encapsulation is one coverlaying process which utilizes relatively low-melt-point adhesives. The encapsulation process is a modification of the laminating process where the material flows between a heated nip roll and a pressure roll. The registration requirement of the coverlay to the circuit must be fairly forgiving in encapsulation as the films tend to move around when they are under pressure and heat at the same time. The adhesive used in this process must be thermoplastic and have relatively low melt points.

Encapsulating is the second most cost-effective coverlaying process. The drawbacks in this kind of coverlaying process are the registration tolerance of the coverlay pattern to the conductor pattern, the low temperature adhesives that must be used, and the limited capability to achieve full lamination (Fig. 2.13).

2.13.2 Coverlay Films: Platen Pressing

In platen pressing, there is much finer registration. The location of the coverlay to the circuit and the platen pressing can be controlled much better. In platen pressing, thermoset adhesives, such as epoxys, phenolics, and acrylics, are normally used. Platen presses are large machines that are heated to high temperatures and used to press materials together. Platen presses have a number of pairs of thick steel plates. Each pair defines a press opening. Into each opening, a circuit layup is placed, consisting of a metal plate, a pressure pad, circuits, and then another plate. A layer of conformal material is added allowing the pressure to be transmitted onto the circuit evenly. Each press is heated with

Figure 2.13 Coverlaying is sometimes done on an encapsulator, a machine which runs the material between heat and pressure rollers simultaneously. (Courtesy of Sheldahl, Inc., Northfield, Minn.)

steam or electricity up to the temperature required to reflow the adhesive with hydraulic pressure available to 200 lb/inch^2. A typical press cycle will run about 1 1/2 hours, including the heat-up and cool-down times.

The best and most expensive coverlaying process is the platen-pressed film coverlay. This process results in definable electrical characteristics and good registration of the coverlay to the copper pattern. It allows for the use of thermoset adhesives which are very temperature-resistant. Each application.s needs should be evaluated to determine which materials and/or processes should be used (Fig. 2.14).

Figure 2.14 Platen presses are heated to high temperatures and utilized to press materials together. (Courtesy of Sheldahl, Inc., Northfield, Minn.)

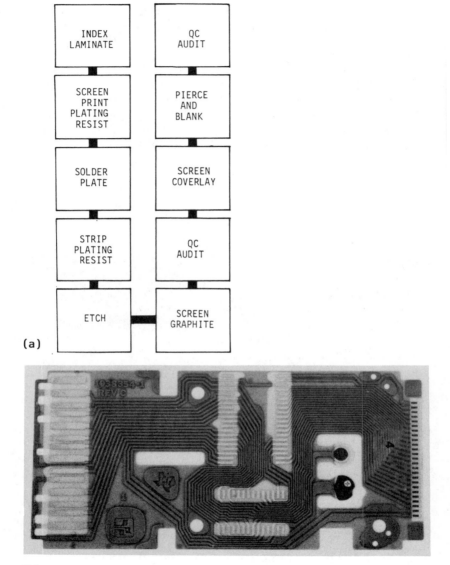

(a)

(b)

Figure 2.15 Print, plate, and etch with coverlay and graphite. (a) Flowchart, (b) model. (Courtesy of Sheldahl, Inc.,
Northfield, Minn.)

2.13.3 UV Curable Epoxy Resins

Screen-printable cover coats are the most cost-effective coverlaying material. Ultrav olet curable cover coats are the
most widely used, but they do not h ve equivalent electrical
properties to films. This coverlaying process tends to be
used as a solder stop, which will protect against a solder
bath, or as an environmental cover. The process involves
screen printing onto an irregular surface, that being the
conductor pattern. The material coats the conductors and
gives a moderately pinhole-free insulating surface (Fig. 2.15).

2.14 DIMENSIONAL FEATURES

When films are used in the coverlaying process, they may be
drilled or pierced. If they are pierced, tooling is necessary.
The nature of the pierce tool is such that the mechanical features of the film are very predictable, and have very little
variance in terms of location and diameter. But this type of
tooling is rather expensive (Fig. 2.16).

2.14.1 Drilling

Drilling of the coverlays is relatively inexpensive from a
tooling standpoint, but it costs more to drill the same hole
over and over again then it does to pierce it. In the ultraviolet curable epoxy resins, the tooling consists of an additional artwork pattern. Generally, a film negative is used
that has the through hole centered on it and a pad to define
the amount of the solderable area that is to be exposed.
Screening a coverlay is cost effective, both from a piece part
standpoint and also from a front-end tooling standpoint.

Pierced films are the least cost effective for small volumes,
but may be applicable when working with very high volumes.
In large quantities, tooling amortization becomes a small portion of the cost of the coverlay feature.

Drilling versus piercing is but one consideration used in
the manufacturing of flexible circuitry (Fig. 2.17).

Figure 2.16 Pierce-and-blank dies define the dimensional features of the flexible circuit. (Courtesy of Sheldahl, Inc., Northfield, Minn.)

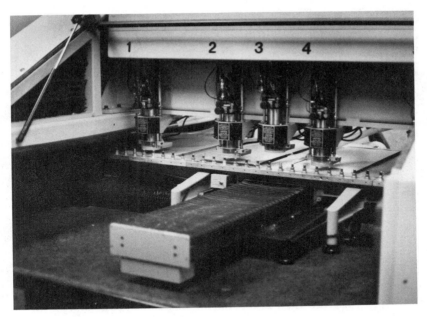

Figure 2.17 The NC driller router is one type of machine which produces small dimensional holes in the flexible circuit material. (Courtesy of Sheldahl, Inc., Northfield, Minn.)

2.14.2 Routing

Routing is utilized when a rigid portion is applied to the flexible circuit. Routin is the cutting away of excess material to define the outline or some parts of the interior of the rigid member of a flexible circuit using an end-mill cutter to mill out the desired shape.

2.14.3 Soft Tools

Rule dies are devices consisting of sharpened steel strips embedded in plywood and operate much like cookie cutters. Rule dies are less expensive and less dimensionally accurate than hard tools. A characteristic of the rule die is that the rules tend to move. When utilizing a rule die, the material inside the cavity tends to push the rules out so that while

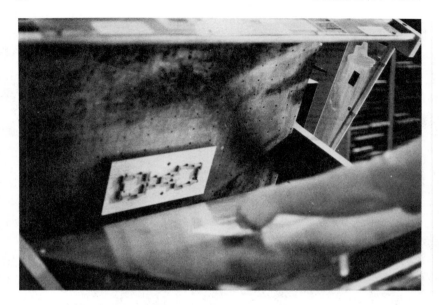

Figure 2.18 Rule dies are used on the Thompson press to cut the outline of the flexible circuit. (Courtesy of Sheldahl, Inc., Northfield, Minn.)

the process begins with an accurate pattern, it becomes progressively farther off datum as the rule die is used. Rule dies can be expected to hold tolerances of ±0.0015 in. (Fig. 2.18).

There are two characteristics that affect the dimensional requirements of the circuit. First is the factor of human error, the ability to set the machine up on target. Secondly, there is process variability. Even the most finely tuned machine process will drift in a predictable way in some range.

2.14.4 Hard Tools (Compound and Progressive)

Hard tools are devices consisting of accurate punches, die plates, and strippers. The tools are fabricated from hardened steel and typically hold tolerances of ±0.001 in. within the die itself. Final dimension tolerances on the circuit vary from ±0.010 to ±0.015 in., due to set-up error on the hard tools and the flimsy nature of the flexible circuit material.

2.15 COMPONENT ASSEMBLY

Flexible circuitry can either be a single component or a mechanism consisting of several components. The most common components added are rigid hardboards and connectors. Standard printed circuit (PC) board material is used, cut to the shape that fits the flexible circuit, and bonded to the circuit with an adhesive. The boarded area, or rigid portion, is used to mount heavy components or to be a part of the general assembly of the module, maybe even a structural portion of the module. A variety of adhesives are used on various portions of the circuits to bond the circuit to a chassis or to itself. The most common kind of hardware added in the component assembly are connectors. Connectors are soldered on, staked on, crimped on, or welded to the circuit. When the assembly is completed, the flexible circuitry can be supplied folded, bent, twisted, or contoured to fit any application. Creasing of flexible circuits, if done, should be carefully evaluated with respect to copper types and sharpness of crease. Creasing can fracture the conductors.

In the component assembly process, bonding of boards can either be done with pressure-sensitive adhesives or thermosetting or thermoplastic adhesives.

2.16 INSPECTION AND TEST

The final step in the production process of flexible printed circuitry is the inspection and testing.

2.16.1 Visual/Dimensional

Visual inspection includes a variety of workmanship requirements such as stains, oxidation and brightness of plating, degree of plating coverage and uniformity, degree of coverlay encapsulation onto the circuit, gel particles in the adhesive, and nonconductive particles in the circuit.

Dimensional inspection includes measurement of the conductor lines and spaces, amount of insulating dielectric between

a conductor and an exterior edge, and the amount of solderable pad around a component mounting hole. Pioneering technology for automated visual and dimensional inspection will
soon be available. Computer-driven cameras which will compare each circuit to a known standard will greatly improve
quality and product reliability.

Today, electrical testing is confined primarily to checking
for shorts and opens. Additional checking can include conductivity, dielectric breakdown, insulation resistance, and
surface resistivity. Most electrical testing for shorts and
opens is done on a 100% basis.

3
Design Considerations

William P. Kelly
Rogers Corporation, Chandler, Arizona

Even the experienced designer occasionally feels as if he has become a novice again when he faces a new packaging method that looks like an old concept but is truly different. Before we begin this chapter on design considerations, we will take the opportunity to address a few of the misconceptions about flexible circuits, which, unfortunately, prevail even after successful application of the technology for more than 15 years.

1. Flexible circuits cost more! Unfortunately they do, at the component level. However, when the concept of installed cost is applied, more than likely you will achieve a cost reduction by employing a flexible circuit. Later in the chapter we will provide an outline of the items to be considered in assessing installed cost.

2. Flexible circuits are just like printed circuit boards! No! They are markedly different. The structural integrity, materials composition, methods of interconnection, methods of handling and assembly, etc. of flexible circuits differ greatly from "hardboards." We are considering two almost totally different concepts for packaging electrical/electronic components. In fact,

one of the few common characteristics is that both are
methods of packaging components.

3.   You cannot replace a wiring harness with a flexible circuit and save money! The oldest high-volume flexible
circuit application belies this statement. It is in use
today after 17 successful years in which it has saved
millions of dollars for its users. It became a success
because the companies that employed the concept in-
vested in the creation of interconnection devices that
were designed specifically for the application of a flexible circuit. It is pictured in Figure 3.1. It is the
automotive instrument cluster circuit.

Hopefully these old prejudices have been dispelled. If so,
we can now move on to consider the task of the packaging
designer who is about to embark on his first flexible circuit
design.

(a)

Figure 3.1 Views of an automobile instrument cluster assembly with flexible circuits. (Courtesy of Sheldahl, Inc.,
Northfield, Minn.)

3.1 APPLICATIONS

There are two basic applications of flexible circuits, those which involve continuous or periodic movement of the circuit as part of end product function, and those which require no movement other than for service needs. These are respectively referred to as dynamic or static applications. The first task of the designer is to determine whether the application involves a static or dynamic mode of operation, selection of materials and method of construction will be dictated by the mode of operation. Figure 3.2 pictures a flexible circuit employed in a disc drive which must survive 400 million flex cycles. In todays' world it is the epitome of dynamic application. Our old friend, the automotive instrument cluster circuit pictured in Figure 3.1 is an excellent example of a static application.

Each type of application, as previously stated, dictates materials and process selection and, as a result, must be carefully considered. The following is a checklist to be considered before beginning a design once the type of application has been determined.

(b)

Figure 3.1 (Continued)

Figure 3.2 A 5 1/4-in disc-drive assembly, using a dynamic
flexible circuit interconnect. (Courtesy of Rogers Corporation, Chandler, Arizona.)

3.1.1 Materials

Static Applications

The focus for static applications should be cost. Consideration must be given, of course, to such items as solderability
if this is a requirement, but the primary emphasis should be
cost. Generally speaking static circuits which can be mechanically assembled will be low in cost if the designer specifies
a polyester or aramid substrate and cover layer and untreated, electrodeposited copper. In choosing substrates the following cost ratios should be used:

Polyester	1	TFE/FEP	8
Aramid	2.5	Polyimide	10
Polyetherimide	4.0	Random polyester/glass	10
		Woven polyester/glass	12

Dynamic Applications

Once a dynamic application has been determined, the broad range of materials available decreases substantially. The TFE and FEP materials disappear because of poor bonds to most adhesive systems and poor creep characteristics. The fibrous materials fall out because the very source of their strength in static applications, the fibers, become an enemy to dynamic survival. The designer's list soon shrinks to four materials: composites, polyesters, polyetherimide, and polyimide. Of the four, only two—polyester and polyimide— have been used to any extent with success.

The sole criterion for selection of a polyester or polyimide material for a dynamic application should be solderability. If the flexible circuit will be exposed to soldering of any form, polyimide materials must be used. If soldering is not a requirement, the designer may choose to save a little money and employ a polyester substrate. The future may determine the viability of polyetherimide materials for a dynamic application, but current data suggest they offer no significant advantage in either performance or cost. The weakest link in the materials chain in a dynamic circuit is the copper employed. In all cases the designer should be sure to specify rolled annealed copper which has been treated for bond enhancement. The specific type of bond enhancement should be selected after careful testing to determine its compatability with the selected adhesive system.

Most adhesive systems will survive well in dynamic applications. A few of the many epoxy systems employed will, over time become brittle and fail in a flexing mode. However these instances are rare indeed and the industry has many systems available for use in dynamic modes.

The many screenable cover coats and solder masks available do not survive well in dynamic applications. These materials provide excellent and inexpensive insulation for static circuits, but should be avoided in dynamic designs.

3.1.2 Circuit Design and Process

Static Applications

The basic rule for designing a static circuit is, "tape the space." In other words, the designer should attempt to maximize the copper left on the circuit once the pattern of the circuit has been etched. This practice lends strength to a continuous web of circuits which minimizes distortion due to roll-to-roll processing. Roll-to-roll processing should be the goal in all static designs, as it is the most economical method of producing flexible circuits. Care must be taken by the designer to insure that the tensions imparted to a web of circuits by the roll-to-roll equipment, combined with the dimensional movement of all flexible substrates due to chemical exposure, do not cause unacceptable dimensional changes.

Dynamic Applications

The primary failure mode of dynamic flexible circuits is conductor cracking due to copper fatigue. This phenomenon can be avoided in most cases by designing the flexible circuit so that the copper is at the neutral axis in all bends. In other words, select adhesive and su strate/cover film thicknesses that insure that the copper is at the center of a cross section of a finished circuit. This may seem impossible in double-sided constructions, however, it may be possible to design the circuit so that the dynamic portion is single-sided. If the designer cannot create a single-sided section he or she can allow a generous flex radius to minimize copper fatigue. Dynamic flexing should not be attempted with multilayer flexible circuits.

Since copper fatigue is the primary problem in dynamic applications, it is logical to remove as much copper as possible from the finished circuit. The rule of circuit design changes to "tape the line" in dynamic applications. Line widths of minimum dimensions (within process capabilities, of course) should be employed and all extraneous copper should be removed.

With most of the copper removed from the dynamic circuit to enhance its survival designers must direct their efforts to sheet processing. Sheet processing will impart fewer stresses to the in-process circuit and limit distortion of the circuit substrate to the very predictable variations due to chemical exposure.

Preparation of interface areas on the dynamic circuit also require special attemtion. Secondary plating/coating of the interface areas, should be specified. Special attention should be paid to keeping the dynamic portion free from all plating/ coating. The roll-solder process described in Chapter 2 is highly recommended, as it allows coating of the interface areas without exposure of the circuit to the harsh chemical environment encountered in electroplating.

3.1.3 Tooling

Static Applications

Tooling for a static circuit should be selected with an eye toward tolerances, production volumes, and cost. Most static circuits do not have requirements for cosmetic edges and so may be readily produced using either steel rule dies or low-cost hard tools employing a hard punch/soft plate concept. Generally speaking numerically controlled (NC) drilling of circuit holes is economical in low and moderate production volumes but is impractical if roll-to-roll processing is used. Full hard tooling, employing both a hard punch and die plate is rarely required in a static application.

A good general rule for selecting a tooling approach for an individual application is to hard tool all applications whose annual production volume will equal or exceed 30,000 circuits (the monthly volume is 2500 circuits). The hard tooling approach will also allow roll-to-roll processing. For circuits whose annual production volumes are below 30,000 pieces it is generally more economical to tool using sheet processing, NC drilling, and steel rule die blanking.

The one exception to the general rule is the tightly toleranced circuit. Whenever hole tolerances are smaller than ±0.003 in. or cutline tolerances are smaller than ±0.015 in. or image-to-blank tolerances are smaller than ±0.020 in., hard tools may be dictated. The designers task is to insure the greatest available tolerance in order to minimize cost.

Dynamic Applications

There are few, if any, dynamic circuits which do not require hard tooling. These are two design constraints which dictate hard tools for dynamic designs. The first is a basic weakness in the two film types capable of being used in dynamic circuits (i.e., poor tear propagation characteristics).

Any minute nick in the edge of a dynamic circuit can become
the starting point for a tear that will result in the demise of
the circuit. The only means of avoiding this possibility is
the use of fully hard tools which have precise die clearance
to insure the highest quality edges. The other important
constraint is the operating environment of the end product
in which a dynamic circuit is used. Many dynamic applications are components of very precise electromechanical devices (such as disc drives), whose operating environment
must be ultraclean. As a result, circuit edges must be totally free from fragments and whiskers that accompany rule die
and hard punch/soft plate blanking. Again, the fully hard
tool with its precise die clearance is the only way to avoid
fragment and whisker problems.

Static applications require few additional considerations in
design. Dynamic applications, however, must consider the
stresses imparted to the circuit in application and a careful
designer will make sure to allow generous bend radii and sufficient travel to provide for the inaccuracies of the flexing
mechanism; and enough freedom of movement to prevent tensile stresses in the dynamic circuit and compensate for multi-
planar stress patterns. A good rule-of-thumb is to allow a
minimum flex radius of 25 tim s the total circuit thickness
(i.e., if T = 0.014, R = 0.350 min). Tensile stresses should
be avoided at all costs since tensile stress in a dynamic circuit can only exacerbate the problem of copper fatigue. A
generous loop allowance will, in most cases, avoid the problem of imparting tension to the dynamic circuit.

One final caution in dynamic circuit design. If it is necessary to mount components on a dynamic circuit, the designer
must be very careful to adequately reinforce those areas containing components so that they are not subjected to stresses
as the circuit performs its dynamic task. The added mass of
components can drastically affect the imparted stresses and
strains to the flexing portion of the circuit and the working
of solder joints will cause those joints to fail. Figure 3.3 il-
lustrates several methods of circuit reinforcement. Reinforcement can be a simple piece of unclad dielectric material or a
complex cast, formed, or machined metal component. The
key is eliminating the stress points at component/circuit
interfaces.

Figure 3.3 Flexible circuits with reinforcement. (Courtesy of Rogers Corporation, Chandler, Arizona.)

3.2 SIZE

Flexible circuits have ranged in size from the circuit shown
in Figure 3.4, to circuits measuring over 100 feet in length.
The larger circuits are generally favored with a repeating
circuit image which can be applied using screen printing or
photoexposing on a roll-to-roll production line. In very large
circuits, the designer must allow for the inaccuracies of a
stepped-imaging process and must take care to design wide
conductors to allow for step misalignment in the transverse
direction and errors in chemical processing.

Generally speaking the designer is not faced with a 100 ft
circuit, but must be careful to design a circuit that is relatively large so that it will process without problems. Typically, problems arise in three areas for large circuits;
plating, imaging, and blanking.

3.2.1 Plating

The typical flexible circuit manufacturer operates a sheet-
plating line, although a few have made the investment in a
roll-to-roll copper- and solder-plating process. Typically
sheet plating places a constraint on the designer since he
must size a circuit to fit a particular sheet size (usually 12 ×
18 inches) in order to have a number of eligible suppliers.
This means the circuit with appropriate margin allowances
must fit into a 12 × 18-in. rectangle. Edge margins are usually 1-in. minimum on the 18-in. dimension and 3/4 of an inch
minimum on the 12-in. dimension. The effective circuit area
(a rectangle of 10.5 × 18 in.) limits the circuit size.

It would appear that roll-to-roll plating would provide a
significant advantage for large circuits, particularly in terms
of cost when one imagines the very slow process of sheet
plating a single, large circuit per panel. In fact, roll-to-roll
processing will only be effective for those large circuits that
have been designed with relatively wide conductor paths
(generally 0.025 in. or wider) since production yields are
significantly lower in roll-to-roll processing of fine line circuits. Sheet plating does not apply tensile stresses to a panel of circuits and, because it processes only a few circuits at
a time, can be fine tuned in-process to adjust for minor circuit variations.

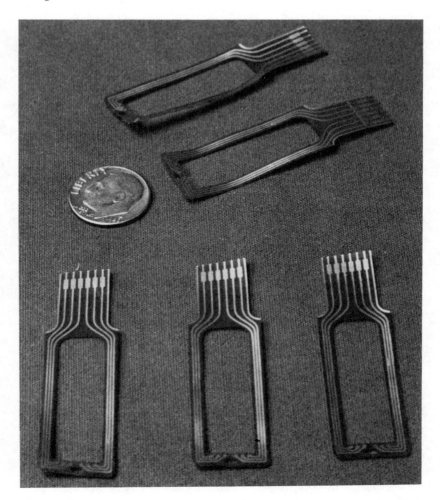

Figure 3.4 A computer memory "crystal" circuit with a dime for size reference. (Courtesy of Rogers Corporation, Chandler, Arizona.)

3.2.2 Imaging

Most screen-printing equipment manufacturers recommend a process rectangle or "footprint" of 24 × 30 in. maximum. If the designer intends to take advantage of this cost effective process, the size of the circuit must be limited so that the

circuit and appropriate margin allowances will fit into this
footprint. We have already looked at appropriate edge margins for plated circuits in the preceding paragraph. As-
suming that a screened circuit needs no plating and can be
simply screened, etched, cover filmed (or cover coated), a
blanked-edge margin of 3/4 of an inch in the machine direction (with-the-web) will be sufficient. Of course the preceding assumes roll-to-roll processing.

Photoimaging offers fewer problems since the process requires no mechanism movement (such as the squeegee of a
screen printer), but merely requires a panel of circuits to
lay flat under vacuum and be exposed to a special light. A
typical exposer window is 24 × 42 in., but much larger units
for sheet processing have been constructed. Since photoimaging is most often used with circuits which require plating
for component assembly, or circuits which employ plated-
through holes, the size limitations are defined by the plating
process.

3.3 ELECTRICAL REQUIREMENTS

In general, the designer is interested in insulating properties
of the selected dielectric material and the conductivity of the
finished line width of the circuit. The nomographs in Figure
3.5 allow the designer to plot the cross-sectional area in
square mils from line width and copper foil thickness, and
then to plot the allowable continuous current in amperes from
the cross-secitonal area and the allowable thermal rise in degrees Celsius. The rules-of-thumb for conservative design
practice is to allow a maximum rise of 20°C for circuitry which
will use polyester dielectrics or adhesives, and 30°C for circuitry using nonpolyesters. Short current spikes which exceed these thermal limits may be encountered successfully.
The nomographs in Figure 3.5 allow the designer to estimate
current carrying capacity of the circuit with thermal rises in
excess of those allowable continuously. Since copper is the
primary conductive material it is good to restate three basic
properties involved in circuit calculations: its resistivity
(0.0000000159 Ωm), its sheet resistivity (0.52 mΩ/2), and
its temperature coefficient of resistance (0.0039/°C).

	Sample Thickness	Dimensional Change at Elevated Temperature	Tear Strength (gm/mil)	Folding Endurance (Cycles)	Ultimate Elongation (%)	Moisture Absorption (%)	Dielectric Constant (1K Hz)	Dielectric Strength (v)	Dissipation Factor (1K Hz)	Flammability	Service Temperature (° C)
R/2400 Dacron-Epoxy	0.004″	.0005 in./in.	40	50,000	15	1	3.2	3,100	0.015	94V-0	-60 to 150
Kapton	0.001″	.0015 in./in.	8	10,000	70	3	3.5	7,000	0.003	94V-0	-250 to 250
Nomex	0.002″	—	49	5,000	10	5	2.0	600	0.007	94V-0	-60 to 120
Mylar	0.001″	.025 in./in.	15	14,000	100	.01	3.2	7,000	0.005	burns	-60 to 95

(a)

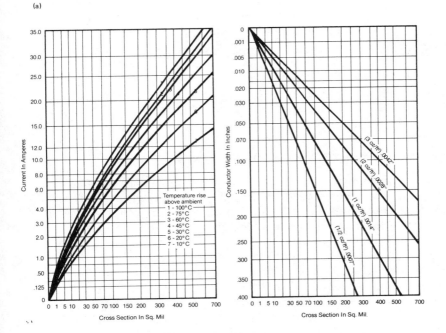

Figure 3.5 (a) Copper conductor design chart, (b) parameters of insulation materials. (Courtesy of Rogers Corporation, Chandler, Arizona.)

In considering selection of dielectrics three properties are usually the designer's focus. These are dielectric strength, dielectric constant, and dissipation factor. Listed below are typical values for the most frequently used flexible circuit materials:

Material (generic name)	Sample thickness (in.)	Dielectric strength (V)	Dielectric constant (1 KHz)	Dissipation factor (1 KHz)
Polyester	0.001	7000	3.2	0.005
Polyimide	0.001	7000	3.5	0.003
Aramid	0.002	600	2.0	0.007
Dacron/epoxy	0.004	3100	3.2	0.015

3.4 ENVIRONMENTAL CONSIDERATIONS

Since flexible circuits are not only a very efficient means of packaging electronic components, but also a packaging system that conserves weight and space, they are often employed in military and aerospace electronics devices. From these end uses comes exposure to a vareity of environmental tests unique to military/aerospace procurement. A designer of the most commercial grade of circuitry should be aware of the component and test specifications published by the U.S. Department of Defense for two reasons. First, most designers have little time to prepare appropriate test programs, so the tests designed and used by the DOD and other federal agencies provide the details of a ready-made test program. Finally, most independant testing laboratories are prepared and familiarized to conduct testing to military standards. Most of the environmental tests contained in the two key Military Testing Specifications (MIL-STD-202 and MIL-STD-810) have been adopted into test programs to evaluate a broad range of commercial/consumer products (everything from computers to home laundry equipment). The flexible circuit designer should also become familiar with Institute for Interconnecting and Packaging Electronic Circuits (IPC) Standards since these standards are becoming industry standards and are being recognized by the American National Standards Institute (ANSI). A good basic standards library for designer reference should include:

General Specifications
  MIL-STD-429   *Printed wiring and printed circuits terms and definitions*
  IPC-T-50   *Terms and definitions*
Material Specifications
  MIL-M-55627   *Materials for flexible printed wiring*
  IPC-FC-231   *Flexible bare dielectrics for use in flexible printed wiring*
  IPC-FC-232   *Specifications for adhesive coated dielectrics films for use as cover sheets for flexible printed wiring*
  IPC-FC-241   *Metal-clad flexible dielectrics for use in fabrication of flexible printed wiring*
General Circuitry Specifications
  MIL-P-55110   *Printed wiring boards*
  MIL-P-50884   *Printed wiring, flexible*
  IPC-FC-240   *Specification for single-sided flexible printed wiring*
  IPC-FC-250   *Specification for double sided flexible wiring with interconnections*
Testing Specifications
  MIL-STD-202   *Test method for electronic and electrical component parts*
  MIL-STD-454   *Standard general requirements for electronic equipment*
  MIL-STD-810   *Environmental test methods*

Armed with these documents the designer has all that is necessary to specify appropriate environmental tests for a flexible circuit. The only remaining task is to select values for the appropriate test parameters (i.e., temperature, relative humidity, exposure time, etc.) that will provide a look into the future in terms of performance while making sure that the testing was not, as is often the case, overkill, which can only lead to overspecification and added cost. There are many pitfalls in environmental testing. The following are suggestions to help the designer avoid those pitfalls.

Thermal Shock Testing

Using the above referenced standards as a guide, make sure the temperatures tested realistically reflect the environment to be seen by the circuit in application. Above all, avoid the temptation to add "20 degrees or so" as a safety factor. You may wind up degrading your flexible adhesive system and failing a circuit that would have performed beautifully in its *real* environment.

Thermal Cycling

Make sure the test temperatures are realistic. It is a relatively common occurrence for designers to specify a totally unrealistic environment in an attempt to compress a real lifetime into a short period. A typical test program involves thermal cycling between - 40°C and +85°C with 30-minute dwell times and 10-minute transition times. It is obvious that the test environment does not exist begging the question, "What have we discovered from this test"? Hopefully designers using this book will specify environments from which something concrete can be learned.

Vibration Testing

As much as designers tend to overspecify the thermal cycling test, they tend to underspecify or, worse yet, fail to specify vibration testing. Since most of the commonly used flexible dielectrics have rather poor initial tear and tear propagation characteristics, the most common form of circuit failure is a tear that propagates and destroys a component solder joint. Vibration testing, especially at low frequencies, can point up potential problems to designers early in the game before the product is in the field and its integrity is open to question.

Environmental Testing

There is a standard group of environmental tests which should be run to qualify particular suppliers of flexible circuits. Once the suppliers have demonstrated their capability of combining and processing the specified materials into acceptable circuits, the testing can be reduced to audit levels.

These are:

Test	Specification
Thermal shock	MIL-STD-202, Method 107
Electromigration	MIL-STD-202, Method 106
Corrosion	MIL-STD-202, Method 101
Vibration	MIL-STD-202, Method 201
Shock	MIL-STD-202, Method 202
Resistance to soldering heat	MIL-STD-202, Method 210
Resistance to solvents	MIL-STD-202, Method 215
Fungus	MIL-STD-810, Method 508.1

In addition to the tests listed above, the designer should develop a realistic set of thermal cycle test parameters that will accurately reflect the temperature extremes, cycle times, and relative humidity levels the circuit will encounter in its application. Attention to acid fog testing and atmospheric pollutants (e.g., hydrogen sulfide) testing should be given when exposures to these substances are part of the application environment.

Dynamic Life Testing

Dynamic circuits should be evaluated for flex life on a test fixture that accurately simulates the flexing mode in operation. Figure 3.6 shows one such test fixture. These fixtures will vary in design to simulate the actual application as will the actual flex test program. Chapter 8 will offer some more specific information on the testing of dynamic flexible circuits.

3.5 DOCUMENTATION

Documentation of flexible circuits is most probably the primary contribution of the designer to the production process. There are almost as many methods of documentation as there are designers to perform a documentation task. It is not the purpose of this chapter to endorse any system of documentation, or, worse yet, to attempt to impose the author's favorite system as the "only" way to get the job done. What are offered are a few general principles that will, hopefully, help the designer avoid some of the pitfalls the author and many others have experienced over the years.

Figure 3.6 A flex tester in operation. (Courtesy of Rogers Corporation, Chandler, Arizona.)

3.5.1 *Forget the Toolmaker, Remember the Inspector*

Almost every drawing in industry is loaded with dimensional references to centerlines, intersection points, reference targets not part of the finished product, etc., etc., etc. The result of this approach to drawing preparation can be seen in the quality area as a multitude of inspectors are constantly employed with the task of determining where, in relationship to the finished circuit, the imaginary points and/or lines fall. The inspection task must be repeated over and over since each part possesses slightly different edges, holes, etc. from which the imaginary datums are determined. The cost impact is obvious both in man hours and dollars.

Most systems of dimensioning are directed toward the builder of tools. It is wise to remember that the toolmaker has but one tool to build, while the inspector has many parts to check for conformance. Anything the designer can do to allow the use of gauges and other inspection tools and avoid the cost and time consumption of slow measuring activity will have a very positive bottomline impact.

A recommended starging point or introduction to the concept of geometric-form tolerancing which allows easy implementation of gauging and fixturing for inspection purposes, is a study of American National Standards Institute (ANSI) Y14.5-1973.

The few minutes spent by the experienced toolmaker in finding the appropriate centers for tool building will be more than adequately compensated by the many hours saved inspection time, which brings us to our next principle.

3.5.2 *Tolerance Each Dimension*

The author realizes that this announcement may result in a drafting room mutiny, but is willing to face that consequence. The same statement applied to toolmakers also applies to drafters, they only do their job once. The inspector on the other hand has a repetitive task and it pays to save the inspector's time. Each feature of a particular component has a unique set of allowances and tolerances that can be applied consistant with good design practice. To apply an overly tight "block" tolerance to all dimensions costs time and money, and virtually guarantees that the quality control "hold cage" will be filled with perfectly functional parts that have minor dimensional discrepancies.

If a particular feature can function well with a tolerance of ±0.100 then it should not bear a block tolerance of ±0.010. A ±0.100 feature can almost always be given a glance rather than measured precisely, which brings us to another principle.

3.5.3 *Wherever Possible Use a General Specification*

Many flexible circuit features such as line width, space width, edge to conductor criteria, lamination criteria, inclusion (or exclusion) of foreign material, may be covered in a general specification. The use of a general specification allows the application of the same "rules" to many different circuits. The designer should move in the following direction when specifying circuit details, from those items which can be generally specified to those items which can be specified on a maximum/minimum basis, to those very few items which need specific dimensions and of course specific tolerances. By taking some time initially to consider carefully each document detail the designer will save a multitude of hours in production and inspection and will insure the use of virtually every functional circuit.

Hopefully the principle of general specification was well received, which brings us to the next principle.

3.5.4 *When Preparing General Specification Use Industry Standards as a Guide*

The Institute for Interconnecting and Packaging Electronic Circuits (IPC) has written and published specifications and standards covering almost every aspect of flexible circuit design. An excellent starting point for the development of your general specification is the *IPC Technical Manual* which is available for $100 from the Institute. Also useful is the *IPC Printed Wiring Design Guide* available for the same price as the *Technical Manual*. The specifications contained in these manuals have been developed by committees which include users of flexible circuits as well as manufacturers and represent very well what is capable of being produced by most manufacturers of flexible circuits.

And finally, we come to our last general principle.

3.5.5 *Remember That Artwork is Both a Document and a Tool*

Because the preparation of artwork falls either into the engineering area or into the production area most designers have a singular view. Artwork is a document in the sense that it

defines the conductor pattern of a circuit. Artwork is a tool
in the sense that it is used to generate a screen printed or
photoprinted image in the manufacturing process and in the
sense that it is often used as an inspection template. A few
helpful hints concerning artwork:

1. The artwork master (a document) must be kept in a
   well-controlled environment at all times. Photographic
   films are very susceptible to dimensional change from
   minor variations in temperature and humidity and from
   exposure to certain chemicals.

2. Artwork masters which must be exposed to the varying
   and chemically harsh environments of the typical flexible circuit factory must be produced on a stable substrate. There is only one substrate material available
   today which will remain dimensionally stable in the
   chemically harsh, and temperature/humidity variant
   environment of a flexible circuit manufacturing facility.
   That material is glass.

   If the reader doubts the truth of this statement, let
   him pause here to read the following chapter to see the
   tolerances necessary to produce high-yield flexible circuits under normal manufacturing conditions *when the
   artwork is assumed to be perfect!* Imagine what these
   tolerances would be if we assumed an art master variation of 0.0005 in./in., which is not at all an unusual
   variation when photographic film is exposed to a factory environment.

3. Wherever possible prepare art masters at an expanded
   scale to minimize dimensional discrepancies. The scale
   should be as large as possible.

4. Remember that most flexible circuits will survive the
   manufacturing process much better if a maximum amount
   of copper is left on the circuit (this rule-of-thumb does
   not apply to dynamic circuits). This "quirk" of flexible
   circuits will affect the designer using a computer-
   aided design (CAD) system which has not been specifically programmed to maximize copper in a circuit
   pattern.

3.6 COST

The designer must be keenly aware of the difference between price and cost. Price is the money paid to procure a component or subassembly. The cost of that component or subassembly will include many other items such as shipping charges, labor, tooling, warehousing, administration, inventory and production control, inspection and quality engineering, rework, reliability administration and testing, documentation, and so on. To merely compare the prices of two alternative packaging systems may be a totally improper and inadequate method of determining their true relative cost.

The best example is perhaps a comparison between a flexible circuit and a packaging system involving two circuit boards and a wriing harness constructed from two connectors and 20 feet of wire of 8 color codes. The flexible circuit arrives at the assembly line as a single component whereas the "traditional" packaging system arrives as two circuit boards and a wriing harness. A quick comparison of the two concepts reveals many hidden but nonetheless real costs:

Cost item	Flex	PCB
Number of part numbers	1	11
Number of purchase orders	1	4
Number of stockroom slots	1	11
Number of subassemblies	0	1
Number of reqork stations	0	1
Number of electrical interfaces	0	2
Number of engineering drawings	1	5
Number of inspection documents	1	6
Number of vendors to qualify	1	3
Number of tooling packages	1	3

It should be clear to the designer that the apparently "cheaper" PCB method of packaging may not prove to be the lowest *cost* alternative. Since the designer is the first person in the chain of technical events that result in a finished product he or she should be aware of the impact on cost that the selection of a packaging system will bear. Only by appropriate attention to all cost details will the designer successfully complete the creation of a reliable yet economical product.

3.7 CONCLUSION

It is the author's sincere hope to have communicated the necessary information to enable the designer to fearlessly set out on the path to a successful flexible circuit design. One final suggestion is offered. When a problem arises, talk first to your flexible circuit suppliers. Their business is flexible circuits and it is in their best interest to help you solve your problem. A competent supplier will have a staff of well-qualified engineers to assist you in the design of your circuit as well as to professionally address and solve any problem that may arise. Involve suppliers early in your design. Listen to their advice. Accept their counsel. Because they are "in the business" they have an excellent handle on both the technologies and the techniques required to manufacture flexible circuits that will meet your needs.

4
Artwork and Design

Ray D. Greenway
Circuit Materials Division, Rogers Corporation, Chandler, Arizona

4.1 INTRODUCTION

Photographic films used for photoresist image placement or screen printing are commonly called photo tools in the printed circuit industry. Whether for hardboards or flexible circuits, the steps required to preapre photo tools are essentially the same. However, there are important differences that introduce some unique requirements in designing and preparing artwork and film for flexible circuits.

4.2 ARTWORK DESIGN

An artwork design print is analogous to a tool design print normally provided to the toolmaker for constructing dies. If an artwork design print is to be prepared it must contain certain information unique to artwork that is not normally found on a part print or die print. Most important are:

1. Photo tool conductor width. The flex circuit manufacturer will often add an etch factor which makes the width of conductors on photo tools wider than the conductor will be on the etched part.

2. Absolute minimum spacing in those areas where con-
    ductors are crowded into a dimension that violates
    standard design rules. Normally the minimum is
    0.005 inch.
3. Radius and center location for all arcs and tooling
    targets or other isolated features.
4. Specification of how conductors are to be blended into
    pads.
5. Border pullback requirements where necessary to in-
    sure copper clearance when blanking, or trim lines
    in the case of prototypes are to be cut by hand.
6. Critical dimensions of features where close tolerances
    are important, as well as some indication of noncriti-
    cal dimensions.
7. Panel layout for those cases where more than one part
    at a time is to be imaged, etched, plated, assembled,
    or blanked.
8. Specification of all logos, part numbers, and special
    coupons that are required for processing or identifi-
    cation.
9. Specification of shrinkage allowance to be built into
    the photo tool and the emulsion direction required.
    Both these are functions of the processes used by the
    flex circuit manufacturer.
10. In the case of double-sided or multilayer circuits, ad-
    ditional information for layer-to-layer registration
    must be provided. Specifically, the layer-to-layer
    tolerance and copper direction for each layer must be
    shown.

4.3 ARTWORK PREPARATION

If an artwork design print is not provided, the artwork specialist must ascertain all the above information and work directly from customer prints and tooling prints. Some artwork
suppliers prepare a layout to the same scale as the artwork
before proceeding. However, in most cases the artwork specialist goes directly from the product engineering data to the
artwork itself. This can present quite a challenge when the
customer print has been drawn by someone not familiar with
some of the difficulties of preparing artwork. For example,

a blend radius may be shown with no indication of the location of its center. Unless the rest of the print is dimensioned from a datum in such a way as to make it easy to locate intersections of projected lines, the artwork specialist is faced with a formidable task trying to locate centers.

Sometimes a product designer will create an impossible geometry such as the example shown in Figure 4.1. The specified width, angle, and end points of the line segments all interact. Before artwork can be prepared, a decision must be made whether to let the width or the angle be the controlling dimension.

A different geometry problem is illustrated in Figure 4.2. Here the problem is one of ambiguity, since there is not way to know the dimensional position of point A. Frequently the geometry problems aren't discovered until the artwork specialist is actually attempting to prepare the artwork. The net result is often wasted time, extra cost, and missed schedules, all of which might have been prevented by better planning.

One final point that is often overlooked by the product designer: It is important to point out all instances where the artwork specialist can use personal judgement and discretion. Overspecifying can be just as costly as underspecifying.

Artwork may be prepared in one of several ways. For example:

1. Hand taping.
2. Rubylith cutting and peeling.
3. A combination of 1 and 2 above with previously prepared film.
4. Computer-aided design (CAD) methods.

Almost all of the above involve working at a magnified scale factor to avoid loss of dimensional tolerance. Hand taping produces crude conductor line edges and should only be used in cases where line quality and dimensional tolerances are not important. Using a 40-mil grid, a good artwork specialist can hold ±10 mil (±0.010 inch) on critical dimensions. If the taping is done at a magnification of 5, this reduces to a tolerance of ±2 mil at ×1. However, seldom is it feasible to work at greater than ×4 and few specialists really hold ±10 mil, so the final artwork will usually have a tolerance of ±5 mil.

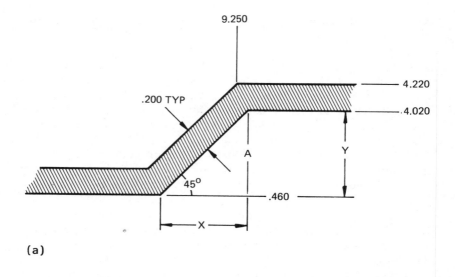

(a)

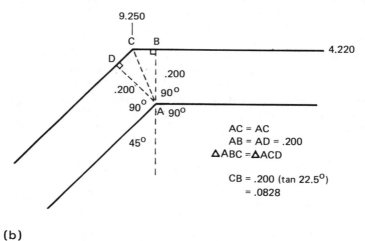

(b)

Figure 4.1 An artwork design print in which the specified width, angle, and end points of the line segments all interact. (a) Calculate A using y = (4.020 − 0.460) = 3.560 and x = y because of 45° angle. Hence A is at 9.31, 4.020. (b) Calculate A using the 9.250, 4.220 dimension. Hence A is at 9.250 + 0.0828 = 9.3328, which is a difference of 0.0228 in. from (a). (Courtesy of Rogers Corporation, Chandler, Ariz.)

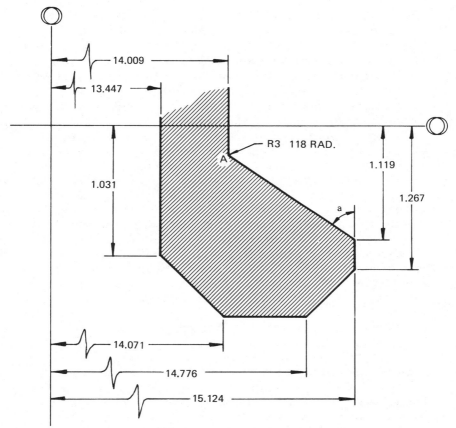

Note that the center of R3 cannot be determined unless point "A"
or angle "a" is specified

Figure 4.2 An artwork design print that is missing vital data.
Note that the center of R_3 cannot be determined unless point
A or angle a is specified. (Courtesy of Rogers Corporation,
Chandler, Ariz.)

Most flexible circuit artwork is prepared by cutting and
peeling rubylith. For precision work, this is best done on a
coordinatograph. However, some artwork specialists are able
to meet close-tolerance requirements by going to magnifica-
tions as high as 25 and cutting with hand tools. The disad-
vantage of this approach is the need to piece together at a

lower magnification since ×25 is too large to cut a complete
circuit on one table. Using a precision coordinatograph it
is reasonable to hold tolerances to ±2 mil. Since most flexible
circuits can be cut at a magnification of 4 or 2, the error at
×1 due to cutting tolerances will be less than ±1 mil for all
except arcs with large radii.

Occasionally, a customer will provide the photographic film
expecting it to be used by the flexible circuit manufacturer.
This poses several problems since the customer film will usually lack etch factor, trim lines or border pullback, and tooling targets. Furthermore, there are frequently cases where
customer film dimensions conflict with customer prints, which
means film, drill tapes, and dies will not match. The art department must make intermediate copies of the customer film
and do a "tape-up" to provide the necessary panel layout and
tooling targets. On a grid this can be held to only ±10 mil.
To obtain closer dimensional tolerance, the artwork specialists often tape a border on the coordinatograph and position
the customer film in the border using the coordinatograph to
measure. Using this method, critical dimensions can be held
to ±2 mil.

Another problem associated with using a tape-up with an
intermediate film is a loss in precision during subsequent
photographic processes. Extra layers of film taped together
increase the risk of distortion and loss of line resolution.

The introduction of computer-aided design (CAD) systems
has solved many of the problems for the flexible circuit artwork specialist. However, CAD has also brought with it
some new constraints for both the flexible circuit product designer and artwork specialist. For example, rounding errors
and calculation variances on source documents accumulate until a discontinuity is discovered somewhere on the part.

To benefit from CAD, the designer must be able to define
every point mathematically. Parts that are not rigorously defined, parts that are drawn freehand, and lines that are
blended to fit, all present problems because of constraints of
the CAD system. Furthermore, a digitizer does not help unless the designer is provided with a scaled model defined on
a reasonable grid.

On the brighter side, CAD systems eliminate the problem
of tolerances in cutting rubylith or doing tape ups. The

designer can manipulate conductors displayed on a graphics terminal until a satisfactory pattern is obtained. Special software packages make it easier to catch many mistakes before committing the artwork to film or glass. However, CAD systems to date have all been directed at the integrated circuit and hardboard industry, not flexible circuits.

4.4 PHOTOREPRODUCTION

The purpose of this section is to acquaint the reader with some variables that affect the cost and time required to obtain photo tools for flexible circuits. Thus far we have discussed only the tolerances involved in preparing a piece of camera-ready artwork or a digital tape for photo plotting.

4.4.1 Camera Reduction

Consider first the reduction and reproduction of a rubylith artwork pattern done at a magnification of 4. For a flexible circuit 18 in. long, the ruby, including alignment targets, will be over 6 feet long. It is nearly impossible to place such a large piece of copy in the copyboard of a process camera without some distortion occurring. The size and weight of the copy tends to cause sagging of the rubylith. Secondly, there are few if any lenses capable of perfect reproduction of such large copy. However, through the skill of the camera person and a certain amount of trial and error, it is possible to produce a ×1 film whose dimensions are within ±1 mil per 10 in. of linear distance. Using photographic film made especially for the electronics industry, it is possible to hold line widths to ±0.5 mil.

4.4.2 Environment

All of the above statements carry certain implications about control of temperature and humidity as well as control of photographic processes. For example, a polyester-based photographic film will grow with either increasing temperature or humidity. An increase in temperature of 7°F or an increase in relative humidity (RH) of 9% will cause an increase in dimension of one mil in ten inches.

4.4.3 Materials

As an alternative to the time and expense of monitoring and controlling film dimensions, the flexible circuit manufacturer might choose to use a different material. Glass photo tools are not affected significantly by temperature or humidity changes. Glass tools are both expensive and fragile. For a multilayer circuit it might cost thousands of dollars to make a minor engineering change on glass phototools instead of hundreds of dollars for film. On the other hand, glass may be the only way to maintain a consistent layer-to-layer registration for close-tolerance circuits.

4.4.4 Step-and-Repeat

A process known as step-and-repeat is used to obtain multiple images on a single photo tool. As in all the other photographic processes, cost is related to precision and tolerance. For example, if the step-and-repeat is done using a CAD system and photo plotting, the cost of photo plotting is proportional to the number of steps. Usually the decision is to start with a photograph of one part and have it stepped on a machine specifically designed for step-and-repeat. To maintain tolerances of ±0.002 inches requires careful setup, since it is easy to have a slight skew when aligning the film in the step-and-repeat chase.

4.4.5 Touch Up

The touch-up process is often called "opaqueing" in the graphics industry. Throughout the entire artwork and photographic process everything must be kept spotless. Even with good habits of cleanliness there will be pinholes and other defects in the exposed film. These must be touched up under a microscope or they will appear as defects in the etched copper part. Once again, cost and precision are related. It can literally take hours to touch up a large panel with many conductors. When the conductors are 5 mil wide or less, it takes a skilled person to touch up without causing an out-of-spec condition.

4.5 CONCLUSION

The illustrations and written examples on the previous pages indicate worst-case situations, and are used to point out the kind of drawing, when supplied to the flexible circuit manufacture, that usually causes delay in tooling and manufacture of the flexible circuits themselves. The designer would be well advised to use a well-known industry standard such as the *American National Standard of Dimensioning and Tolerancing* (ANSI Y 14.5).

5
Design Specifications

Printed circuit boards can generally be manufactured with fairly close tolerances. However, flexible circuitry because of the contraction and expansion of base films requires tolerances that are somewhat more relaxed for economical production. The base material expansion and contraction is caused partially by the inherent characteristics of the film itself and partially because of movement due to tension release after etching. The tension which is released is that which is built up between the base material and the copper during lamination process. Because flexible circuits are a yield-oriented product, realistic tolerancing is the best way to reduce flexible circuit costs.

5.1 CUTLINE

Cutline tolerances can be held as tightly as ±0.005 inch if class A hard steel tooling is used (Fig. 5.1). Lower cost precision steel rule dies can be used if cutline tolerances can be relaxed to 0.015 inch.

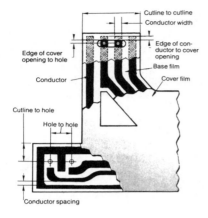

Dimension	Tool	Printing process	Tolerance[a] (in.)
Cutline to cutline	Class A		0.003
	Steel rule		0.015
Cutline to hole	Class A		
	progressive		0.005
	Class A + NC		
	drill		0.010
	Steel rule		0.015
Hole to hole	Class A		0.003
	NC drill		0.005
	Steel rule		0.010
Edge of conductor		Photographic	0.015
to cover opening		Screening	0.020
Edge of cover		Photographic	0.012
opening to hole		Screening	0.015
Conductor spacing		Photographic	0.001
		Screening	0.002

[a] For copper-polyimide flexible circuits dimensions up to 5".

Figure 5.1 Flexible circuit tolerances vary depending on the materials, tools, and processes used. The data listed here are typical for copper-polyimide flexible circuits. Approximately 0.001 in. additional is required for each dimensional inch over 5 in. Random fiber aramid and polyester materials may require up to twice these tolerances.

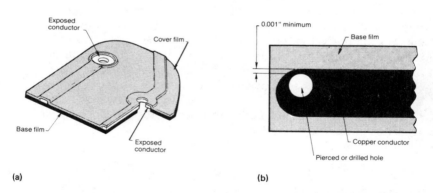

(a)   (b)

Figure 5.2 (a) Conductors on a single-sided flexible circuit can be exposed for termination through the base film as well as through the cover film. But this construction is costly. A similar result in some cases can be obtained by folding a one-sided circuit. (b) Allowing for all tolerances, holes in copper conductors should leave a minimum 0.001 in. of copper at all points around a hole for satisfactory soldering.

5.2 HOLES

Hole tolerances can be held to ±0.003 in. using class A tooling punches, but must be relaxed to 0.005 in. when using
numerically controlled drills (Fig. 5.2). Holes which are
somewhat larger can be punched with tooling made from
steel-rule dies but the tolerances from hole to hole in this
case must be further relaxed to ±0.010 inch.

5.3 PAD SIZE AND LOCATIONS

Pad size is generally determined by the total available real
estate within a flex circuit configuration. It is obvious that
a very small circuit with many, many holes will only allow
small pads. Optimum pad diameters should be three times
the hole diameter but it is possible to tighten these up to a
diameter twice the hole diameter if lack of room dictates this
approach. The reason that the best designs have large pads
around the holes is that larger pads will have a tendency to
resist pad lifting or delamination during soldering operations.
Additionally, larger pad sizes allow for 360° encapsulation or
"capture" of the pad by the cover film insulating layer (Fig.
5.2). On two-sided circuits this is even more critical since
the single hole must match up with an artwork generation on
both sides of the circuit so that the hole has a pad completely
around it on both the top and bottom of the circuit. The
front-to-back registration of pads can be held as tight as
±0.010 inch but ±0.015 inch is preferable (Fig. 5.3). If the
pads are too small, then the plated hole, which is also used
as an interconnecting conductor between the front and back
sides of the circuit, will surely break out from the edge of
one or both of the pads. By reviewing carefully the drawings and tolerance charts at the end of this chapter, you will
get a much needed perspective as to the requirements of the
various design considerations. (See Figs. 5.1-5.4.)

5.4 EDGE INSULATION

Edge insulation is a matter for the designer to consider relative to application. If the flexible circuit is used in an application where its edge will not be touching another conductive part then it would sometimes be alright to allow the

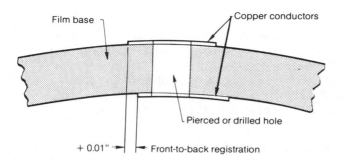

Film base

Copper conductors

Pierced or drilled hole

+ 0.01"   Front-to-back registration

Figure 5.3 When flexible circuit conductors are attached to both sides of a base film by photographic methods, front-to-back registration tolerance typically is ±0.01 in. But registration tolerance is ±0.015 in. where screening processes are used.

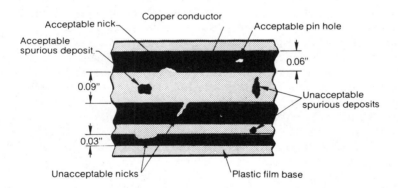

Acceptable nick

Copper conductor

Acceptable pin hole

Acceptable spurious deposit

0.06"

0.09"

Unacceptable spurious deposits

0.03"

Unacceptable nicks

Plastic film base

Figure 5.4 Defects occasionally occur in flexible circuits as a normal result of the manufacturing process. Pinholes or nicks in conductors or spurious copper deposits between conductors are typical defects. When conductor or spacing defects occur in the form of nicks and/or spurious deposits, they should not reduce the dimension by more than 20% of the nominal width.

copper to extend directly to the edge of the circuit. This would, however, also allow the possibility of contamination between the base and the copper or between the cover film and the copper. A better method would be to have 0.010 or 0.015 in. of insulation around the edge of the outer most conductor on the circuit, simply as insurance which would provide insulation integrity, since the cover film would be directly laminated to the base insulator with no copper in between. This design feature is basically one of perimeter tooling and conductor location on the photo tools. Additional information can be derived in the artwork and design consideration chapters.

5.5 CONDUCTOR AND SPACE SIZING

Conductor and space sizing are critical, since extreme minimums of both conductors or spaces should be eliminated because of the possibility of the circuit being lost during the etching process. Allowances should be made for nicks and/or pinholes in the conductors and spurious copper in space locations (Fig. 5.4). The width of the actual space or line itself should be at least 0.020 in. or more if possible, providing a 100% use of the tolerance will not decrease the conductors or the spacing more than 50%.

After the general layout of the circuit has been accomplished, it is a good idea to make a "paper doll" of the circuit with all lines and pad locations drawn in. This paper doll is then bend and/or folded into the proper configuration based on the application. Care should be taken to observe the conductor routing, especially at the point where bends or folds will occur in the application. At these points the conductors should enter and leave the fold line in a vertical orientation or 90° to the fold.

5.6 COVER FILM CONSIDERATIONS

Cover film, is an important feature of a flexible circuit. It insulates and protects the conductors after etching and is usually used to capture hole pads as previously discussed. The registration of cover film opening can be, with respect to the perimeter of the circuit, most economically called out at ±0.020 in. A tolerance of half that or ±0.010 in. can be held with precision tooling and pin registration of the cover film to the base material at the time it is applied and tacked (Figs. 5.1, 5.2). The cover film is then laminated to the circuit. Sometimes designers choose to use the cover film to completely capture each hole pad on the circuit. This is done by drilling or punching individual holes in the cover film for each pad. We call this individual padbearing. A less costly method of allowing the cover film to hold down pads is called group bearing (Fig. 5.1). In this feature, slots are cut in the cover film so that opposing sides of a series of pads in a row are captured. While this does not completely encapture each solder pad with a top insulator, it is a satisfactory method for most applications. The openings themselves can be registered to the hole pad as close as ±0.005 in., but manufacturing-oriented problems such as adhesive squeezeout during the lamination process tend to make a tolerance this tight unrealistic. A more realistic tolerance depending on tooling would be ±0.010 or ±0.020 inch.

ACKNOWLEDGMENT

The author would like to thank Andrea Donahue for the drawings in this chapter.

6
Flexible Circuit Features

There are many features of flexible circuits which relate only to flexible circuits and are not used with other kinds of printed circuitry. Because the circuit has the inherent characteristics of movement and because the insulation is extremely thin, there are many things which can be done during the manufacturing cycle which aid in the final interconnection process itself (Fig. 6.1).

6.1 FOLDING AND FORMING

There are occasions when a circuit needs to be bent into a specific position to fulfill its proper place in the application. The folding and/or forming of flexible circuitry can be accomplished quite easily by hand or with the use of tooling jigs. It is important to note that there is really only one commonly used insulating film that holds a fold permanently. This particular material is polyester and its other characteristics have been discussed in earlier chapters. Since polyimide cannot be used in an application which requires a circuit to be formed into shape and retain that shape through the memory of the film, the memory must then be relied upon by

Figure 6.1 Different kinds of flexible circuits. (Courtesy of Rogers Corporation, Chandler, Arizona.)

the copper foil on the circuit itself. Polyimide cannot be formed in either a cold or heated state. A polyester-based circuit can be formed in a simple way, by bending it into the desired shape, holding it in that shape with appropriate jigging, and immersing the circuit in boiling water for 15-20 sec. After the circuit is removed and cooled, the formed features are held in place by both the memory of the polyester film and the copper foil which is attached to it.

One popular kind of forming which is used, and on which there is a patent held by Sanders Associates, is a reverse coil. In this kind of coil the polyester-based circuit, usually a printer cable or something similar, is wound around a mandrel and then heated as previously described. After this product has been formed it winds and unwinds similar to a window shade action. This particular kind of feature is handy in applications such as printers and in electronic instrumentation, where the flexible circuit can be removed from an instrument for an active or passive component check and then replaced in the instrument without disconnecting the circuit from the instrument itself.

There is another kind of material which is flexible and which can be bent and formed without heating. It is available in dielectric thicknesses of 0.015, 0.020, and 0.030 in., and unlike polyester film which has not had special fire retardants mixed in the adhesive system, it is rated 94VO for self-extinguishing requirements specified by Underwriters Laboratories, Inc. This material is manufactured under the trade name BEND/flex, and is manufactured by Rogers Corporation (Fig. 6.2).

Figure 6.2 Bendable or "flexible" hardboard for static, multiplane interconnects. (Courtesy of Rogers Corporation, Chandler, Arizona.)

6.2 DOUBLE-SIDED ACCESS

A feature which is sometimes known as back-barring pads is used when a mechanical or soldered contact is required on both sides of a single-sided circuit. This feature is manufactured by either mechanically etching or mechanically abrading the insulating film from the copper foil on the base side of the laminate. A better method for accomplishing this is to punch or drill holes in the base film prior to the manufacture of the laminate itself. In this manufacturing step, the base film web is first indexed with holes, which will align both the tooled portions of the circuit as well as its art-tooled image generation on a multiple-pattern basis at predetermined distances along the web. This is known as the image or tooling footprint. After the index punching is accomplished, a special punch or drilling program puts holes in the interior of the web which will end up being the holes in the base material allowing the copper to be exposed through the bottom of the circuit. After this is done, the web is laminated to copper foil as is normally done in the standard flexible circuit material manufacturing process.

6.3 UNSUPPORTED CONDUCTORS

Unsupported conductors are sometimes specified when it is imperative to make contact with the metallic foil itself, but in such applications which will not permit inclusion of insulating film. In this case not only must the cover layer insulation be removed, but the area must have the base insulation absent as well. One of the more popular methods of doing this is to prepunch the base material before lamination as is done in the double-sided access process. The difference is that when the artwork image is generated onto the artwork surface, part of that image is registered on to the area which ha no base supporting material underneath it. In this process the copper is removed from both sides of the circuit during etching which leaves circuit traces or lines with no support either on the top or the bottom (Fig. 6.3). Circuits with this feature are very delicate and require cautious handling as well as packaging and shipping precautions. It is also important to note that this kind of feature along with the double-sided access

Figure 6.3 Circuits with nonsupported conductors. (Courtesy of Rogers Corporation, Chandler, Arizona.)

feature is a fairly important cost component in the manufacture of the flexible circuit itself. Avoid it in your design if you can.

6.4 RIGID REINFORCEMENT

Designers who wish to automatically insert components onto their flex circuits or insert flexible circuits into hardboard-type edge connectors, may require a rigid reinforcement as part of the flexible circuit itself (Fig. 6.4). This is a

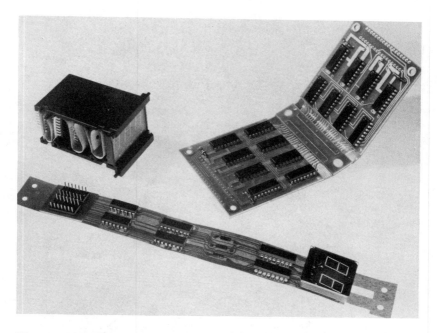

Figure 6.4 Examples of flexible circuits with rigid reinforce-
ments for component mounting and support. (Courtesy of
Rogers Corporation, Chandler, Arizona.)

secondary and somewhat expensive operation since most of
the manufacturing cycle is done by hand. The reinforcement
used is generally an FR4 or glass epoxy hardboard with a
specified thickness for the application. Sometimes a lower
cost material such as a paper phenolic is used. The rigid reinforcement must be cut to conform with the perimeter cut-
lines of the flex circuit, and holes in the reinforcing material
should be at least 0.010-0.012 in. larger than the flex circuit
throughholes which allow for the registration of the circuit

to the hardboard backer. When piercing hardboards, remember to consider the aspect ratio relative to the hole sizes. The hardboard hole diameters should be at least twice as large as its thickness. Reinforcing boards can be attached to flexible circuits with either adhesive systems which require laminating under heat and pressure or with pressure-sensitive adhesive (PSA) applied to the hardboard. Adhesive applied to finished circuitry in a roll-to-roll process, generally produces unsatisfactory results relative to the adhesive transferring to the through holes already punched in the web. Free film adhesive with release material on either side can be formed into panels and prepunched with the hole pattern to allow application to the circuit with proper registration and without the added problems of hole plugging. There are many pressure-sensitive adhesives on the market today which have excellent bonding characteristics, and although they are somewhat more costly, can be more economically used because of the elimination of a hot laminating cycle.

Sometimes reinforcement of a flex circuit is needed, but total rigidity is not desired. In this case, an additional piece of flexible circuit insulating film sometimes thicker than the original circuit film can be easily attached to the flex circuit with either a standard adhesive system or one of the PSA types previously mentioned.

6.5 PLATING

There are many types of plating used in flexible circuitry. The most common type is solder plating. Most solder-plated circuits which have solder over all of the conductors and pads are done electrolytically. This kind of plating is called pattern plating. The process for pattern plating is such that a negative image of etch resist is plated on the copper portion of the laminate and then the circuit is solder plated. The copper foil is the conductive portion and represents the cathode in the tank. Tin/lead annodes will gradually plate onto this copper surface except where the etch resist has been screened or resist imaged. After plating, resist is scrubbed away and the circuit is then etched. The plated electrolytic solder acts as a resist, resulting in a circuit whose traces have solder plate on the top circuit but have exposed copper

on the edges. One of the disadvantages of this particular
process is that undercut produced in the etching process allows slivers of the solder plate to break off and possibly
cause shortening of the conductors. When properly controlled, however, this is a reliable process and has been used
for many millions of circuits particularly in the telecommunications industry. When flexible circuits are specified to have
insulating cover films, it is desirable not to have solder plate
underneath the cover film. The primary reason is that during wave- or hand-soldering operations, solder plating would
melt and the melting would continue underneath the cover film
at the perimeter of the solder pad. This would cause delamination and, in some cases, an unreliable circuit when long
life is required. The most popular method of solder plating
when cover films are used is the individual pad method. This
is not an electrodeposited process but is a mechanical fusing
process which is accomplished in one of several ways. After
the flexible circuit has been manufactured through the cover
layer process, paste-containing solder is screened onto the
individual pads of the circuit. The circuit is then put on a
conveyor belt which runs through a specialized focused infra-
red reflow oven. This melts the solder paste and causes the
solder to fuse with the copper pads. Holes are then punched
in the pad area through both the solder plating and the
copper.
     Another method is a solder-dip process where the entire
circuit is dipped into molten solder and then leveled with a
hot-air leveling system. An alternative method that is often
used is a roller coating method, whereby a roller coated with
molten solder, makes contact with the pad areas of the circuit. Solder thickness can be varied depending on the tension or distance between the upper roller and the lower roller
which is coated with solder. This adjustment varies the pressure applied to the circuit when put through the roller coating process. The action is similar to the old washing machine
wringer used to remove excess water after washing clothes.
Depending on the process used, solder may be placed on circuits with a range of thicknesses varying from 2/10-3/10 of a
mil as applied in the electrolytic process, to as much as
0.002 or 0.003 in. in one of the fusing processes.

Other materials used to plate circuitry, include tin, nickel, and gold. There are several kinds of gold processes used and you should consult your flexible circuit vendor relative to your specific application before making a decision to use one of the many hard or soft gold or gold over nickel platings available. There is some controversy relative to the flexural characteristics of a circuit after a nickel-plating process has been used. Nickel being hard is used to prevent gold migration, however, it is also brittle and could cause conductor cracking when bent or twisted. Several companies have used gold over nickel plating in dynamic applications and have succeeded as long as the nickel plating was extremely thin. Precious metals are generally only specified when absolutely required to keep contact resistance between two mating objects such as between a connector and a circuit, at a very low level. Most flex circuit applications do not require this kind of plating and its associated expense, but use tin or solder plating for both circuit integrity and lower manufacturing costs.

7
Termination of Flexible Circuits

During the 1960s and into the early 1970s, applications for flexible circuits grew slowly, primarily because terminating, or connecting the circuits to adjacent components or other parts of the system was difficult. Few connectors were available to work properly with thin substrates, pressure systems were not generally available, and process engineers were afraid to use solder systems which were destructive to flexible circuits when not set up and operated properly.

Presently, there are many viable systems which can be used with high degrees of efficiency. Costs have been reduced through the use of carefully controlled soldering systems, well-trained personnel in hand-soldering operations, as well as semiautomated connector installations using chain and mass terminated connectors. One other soldering system which works well, but which has cost limitations relative to initial capital expenditures, is vapor-phase soldering.

Some of the more common examples of termination will be examined in this chapter.

7.1 SOLDERING

By far, the least expensive method of device and circuit
termination is through soldering, using semiautomatic equipment. Standard wave-soldering equipment can be used with
proper precautions. It is necessary to reduce the solder
temperature to a low temperature zone between 500 and
520°F. Using solder in this zone, plus increasing belt speed
so that the dwell time is no greater than 5 seconds will provide a good solder joint. Most flexible circuits using polyimide film together with a high temperature epoxy or acrylic
adhesive system will stand up well in this manufacturing system. It is desirable to preheat the circuits immediately prior
to soldering since polyimide film is hygroscopic and it is necessary to drive off moisture before the film comes in contact
with the solder. Failure to do so may result in water vapor
in the film exploding and tearing holes in the circuit itself.
Many production facilities use glass epoxy or stainless steel
jigs which have holes cut in them to allow the solder to come
in contact with only the terminating pads themselves, and not
the entire circuit. This method has been used satisfactorily
with circuits made from random fiber aramid or random fiber
glass-polyester bases as well as polyester film. Most wave-
solderable polyester film circuits are made with film at least
5 mil thick and with 2 oz copper conductors. This construction is still less expensive than other composites or films
which are thinner.

Hand soldering is more expensive than even some connector
systems, but is used for short runs and prototypes when
tooling or small connector component purchases are not
desirable.

7.2 PRESSURE TERMINATION

One method used for termination which requires careful design is with pressure. Spring steel clips can be used to clip
together the flexible circuit contact area with a mating area
on a printed circuit (PC) hardboard. A small notch on either side of the flex circuit which mates to small pins on the
hardboard assures contact alignment prior to installing the
clip.

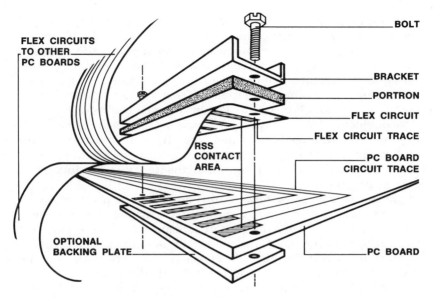

Figure 7.1 A pressure termination system.

Some pressure systems use a nonporous closed cell poly-
urethane material to provide an even flow of pressure across
multiple pads on the end of the flexible circuit. After alignment is assured, a pressure inducement device, such as a
0.060-in. thick steel bar is screwed down. This bar exerts
force against the pressure strip which deforms slightly to allow equal force to be exerted on each conductor pad. The
lack of "memory set" in the pressure material insures continued pressure at the pad locations for many years.
   Figure 7.1 shows this system.

7.3 CONNECTORS

By far the widest used terminating method for flexible circuits is the connector. There are many different types and
styles. Some of these will be described briefly, and then
some examples of common usage among products by four different manufacturers will be examined.

The individual connector can be attached to flexible circuit traces line by line with a hand-crimping tool. This is a satisfactory approach for prototypes or small production runs. In the case of heavier production quantities, a semiautomatic terminating machine may be used with individual connectors connected in a chain and supplied from a spool. This is much faster and is lower in cost. Faster yet are the mass termination devices. These devices provide terminating pins on a row of pads and are all crimped into the circuit with only one or two strokes of a specially tooled press. Using these systems, as many as 32 pins can be connected to a flexible circuit at one time.

There are also mass terminating devices which I call passive because they are not permanently crimped onto the flexible circuit. Instead, the circuit is manufactured in such a way as to allow conductors to come to one end of the circuit, ending in contact "fingers." These fingers are then plugged into the connector which applies individual contact force through a system using a single or dual cantilever spring with single or bifurcated contacts.

7.4 CONNECTOR SOURCES

The following examples represent some of the many choices of connector manufacturers and systems available for use with flexible circuits.

7.4.1 AMP Inc.

This company manufactures many connectors which provide proper interfaces with flexible circuitry. One of the easiest types to use is a one-piece device which incorporates the terminals and housing. It has solder tails for wave-soldering into the PC hardboard to which the flexible circuit is to be connected. (Figs. 7.2 and 7.3.)

Another type of device is not really a connector, but a nonelectrical spring steel device which presses the flexible circuit pads down to the mating pads of a p inted circuit hardboard. This product is called the zero insertion force (ZIF) connector (Figs. 7.4 and 7.5).

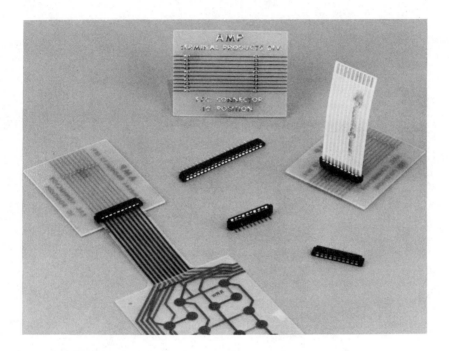

Figure 7.2 Trio-Mate 57287. (Courtesy of AMP, Inc., Harrisburg, Pa.)

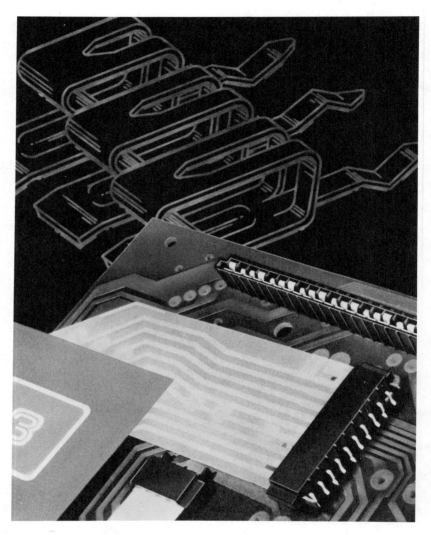

Figure 7.3 Trio-Mate 62552. (Courtesy of AMP, Inc.,
Harrisburg, Pa.)

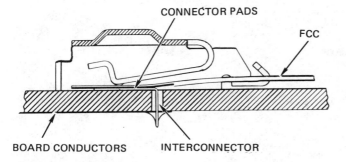

Figure 7.4 ZIF connector diagram. (Courtesy of AMP, Inc., Harrisburg, Pa.)

7.4.2 Berg Electronics

A division of du Pont, Berg Electronics has a broad array of connectors which meet most interconnecting requirements. One type is a mass terminated device, meaning that its individual terminals are all crimped into the multiple conductors of the flexible circuit at the same time using a semiautomatic or hand tool (Figs. 7.6-7.9).

7.4.3 Burndy Corporation

A trademark device called Flexlok, FC series connector is manufactured by Burndy Corporation. This device is a self-contained connector using terminals incorporated in a housing which supports the terminal array. The terminals in this connector are designed to provide a gastight connection based on a patented principle of plastic deformation of the interacting metals to break down oxide surfaces. The points, or contact areas, actually dig into the copper traces or pads on the flexible circuit. Although this interconnect is good for low-level circuit signals, the number of insertions and withdrawals is limited to a half-dozen or so because of the abrading effect on the circuit traces. Figures 7.10 and 7.11 show this kind of connector along with a typical manufacturer's description and specifications.

Figure 7.5 Low-cost ZIF connector for flat conductor cables.
(Courtesy of AMP, Inc., Harrisburg, Pa.)

Figure 7.6 (facing page) (a) Berg solder-tab Clincher connector mass terminates flat-conductor, flat-cable, or flex circuitry in just 10 seconds. (b) Solder-tab Clincher about to
mate with Berg Minisert low-profile socket. The solder tabs
can also be soldered directly to the board. (Courtesy of
Berg Electronics, a division of E. I. du Pont de Nemours,
Camp Hill, Pa.)

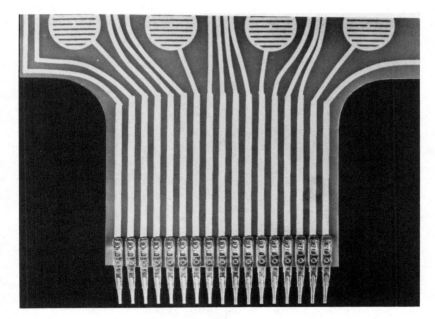

(a)

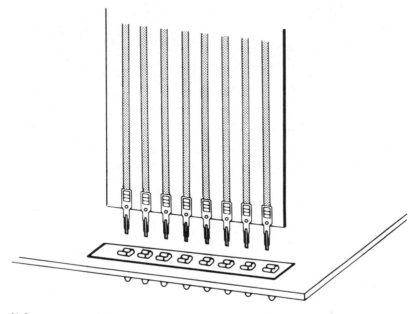

(b)

Figure 7.7 Typical flexible circuit connectors. (Courtesy of Berg Electronics, a division of E. I. du Pont de Nemours, Camp Hill, Pa.)

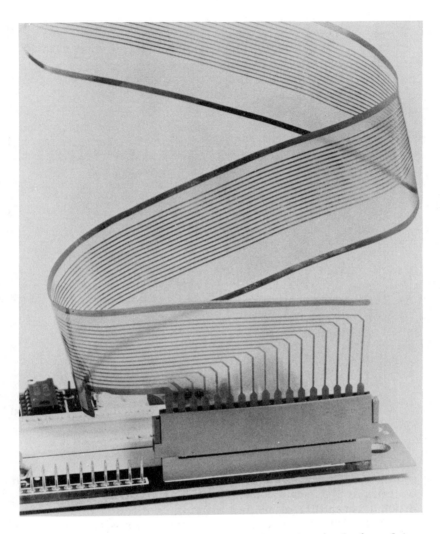

Figure 7.8 Berg's Clincher connector system is designed to mass terminate flat conductor, flat cable, or flex circuitry. It is used in packaging applications requiring maximum flexibility, optimum space, and weight savings. The Clincher meets the industry's need for an economical yet reliable termination method for these applications. (Courtesy of Berg Electronics, a division of E. I. du Pont de Nemours, Camp Hill, Pa.)

Figure 7.9 Berg's male Clincher offers the same features and
benefits as its female counterpart, but gives its user the additional versatility of making a flying connection to a female
Clincher, Berg PV housing assembly, or any other receptacle
which can mate to 0.025 in. square pins. (Courtesy of Berg
Electronics, a division of E. I. du Pont de Nemours, Camp
Hill, Pa.)

- Eliminates use of gold in circuitry for savings up to 60%
- High reliability even in adverse environments
- Soldered to P. C. board. No separate handling
- Supplied with contacts assembled. Eliminates need for special tooling or operator training
- Compact, low-profile design for higher densities
- UL 94 V-O flammability rating

Figure 7.10 Flexlok FC series connectors are specially designed GTH connectors for use with flat conductor cable and flexible printed circuits. (Courtesy of Burndy Corporation, Norwalk, Conn.)

Catalog Numbers		Dimensions in Inches		
Top Entry	Side Entry	A	B	C min
HBLB6S1	HBLB6R1	.820	6	.706
HBLB7S1	HBLB7R1	.920	7	.806
HBLB8S1	HBLB8R1	1.020	8	.906
HBLB9S1	HBLB9R1	1.120	9	1.006
HBLB10S1	HBLB10R1	1.220	10	1.106
HBLB11S1	HBLB11R1	1.320	11	1.206
HBLB12S1	HBLB12R1	1.420	12	1.306
HBLB13S1	HBLB13R1	1.520	13	1.406
HBLB14S1	HBLB14R1	1.620	14	1.506
HBLB15S1	HBLB15R1	1.720	15	1.606
HBLB16S1	HBLB16R1	1.820	16	1.706
HBLB17S1	HBLB17R1	1.920	17	1.806
HBLB18S1	HBLB18R1	2.020	18	1.906
HBLB19S1	HBLB19R1	2.120	19	2.006
HBLB20S1	HBLB20R1	2.220	20	2.106
HBLB21S1	HBLB21R1	2.320	21	2.206

PERFORMANCE CHARACTERISTICS

Dielectric Withstanding Voltage: 500 Volts A.C. min

Current Rating: 3 Amperes max.

Operating Temperature: -55°C to +75°C

Contact Resistance: 15 Milliohms max. (initial)

Insulation Resistance (500 V. D.C.): 5000 Megohms min

Vibration: MIL-STD-202, Method 204, Condition D; no electrical interruption greater than 1 microsecond

Mechanical Shock: MIL-STD-202, Method 213, Condition D; no electrical interruption greater than 1 microsecond

Moisture Resistance: MIL-STD-202, Method 106, omit 7a and 7b; 100 Megohms minimum

Thermal Shock: MIL-STD-202, Method 107; -55°C to +120°C

Durability: 5 Cycles min

Gas Tightness: (Nitric acid fumes followed by ammonium sulfide fumes) Contact resistance 25 Milliohms max.

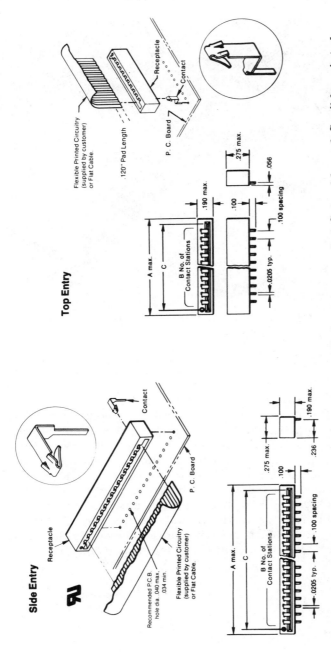

Figure 7.11 Burndy sales drawing (SD 77520) for recommended flat cable and flexible printed circuit preparation for use with Flexlok FC Series Connectors. (Courtesy of Burndy Corporation, Norwalk, Conn.)

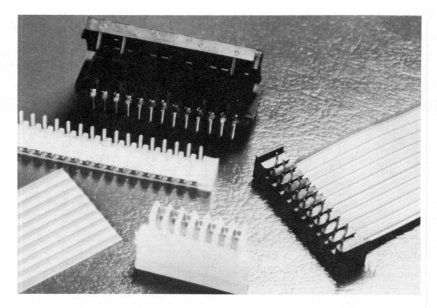

Figure 7.12 Molex 4850 Connector. (Courtesy of Molex, Inc.,
Lisle, Ill.)

7.4.4 Molex Inc.

Molex Inc. manufactures a line of connectors which can be
used with flexible circuits and flat cable. One style is a zero
insertion force policized device, which locks the flexible circuit down against the connector terminals. The other sides
of each terminal in the connector are formed into solder tails
which are inserted into and then wave- or hand-soldered to
the printed circuit hardboard (Fig. 7.12). Other connectors
available include the 7583-CN, with low insertion force, which
contacts exposed pads on the flexible circuit (Fig. 7.13).
Another low-profile, low-insertion force device, is the 5229
series, which also mounts on a printed circuit hardboard
(Fig. 7.14).

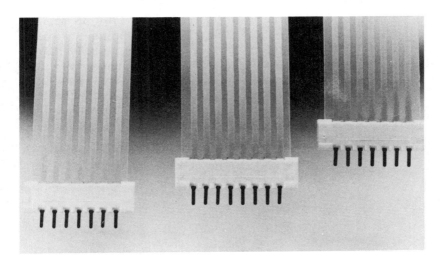

Figure 7.13 Molex 7583-CN Connector. (Courtesy of Molex, Inc., Lisle, Ill.)

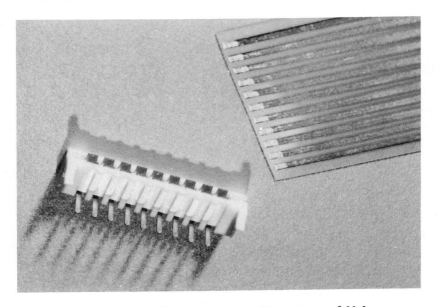

Figure 7.14 Molex 5229 Connector. (Courtesy of Molex, Inc., Lisle, Ill.)

The connectors discussed in this chapter are by no means the only choices the designer has when terminating flexible circuits. There are a score of companies with literally hundreds of designs which can fit many different design requirements. Some connectors are simple, low cost, and need little or no tooling for their use. Other designs require sophisticated tools or capital equipment to install them properly. Advice can be obtained from both the flexible circuit and connector vendors.

8
Testing and Inspection

Although testing and inspection of flexible circuits fall into two separate categories, many of the activities related to the two subjects are the same.

As stated in previous chapters, it is important to have a properly designed product both in terms of master artwork and dimensional drawings, fully toleranced.

The simplest and lowest cost method of inspecting finished flexible circuits is by eye, without magnification. It is obvious that not all designs will fall into this category since closely toleranced, dense circuits preclude visual inspection with the naked eye. Circuits with wide lines and spaces, along with reasonable tolerances can be visually inspected by eye either in panel form, or in a roll-to-roll configuration, over a specially set-up light table. This kind of light table is usually equipped with an operator-controlled speed device which allows for variable rates of inspection speed, depending on the defect nature and frequency.

Circuit designs with very fine lines and spaced and very close tolerances are often inspected under magnification. Many of these circuits fall into what is known as "computer-grade" circuits. This kind of circuit can be very dense,

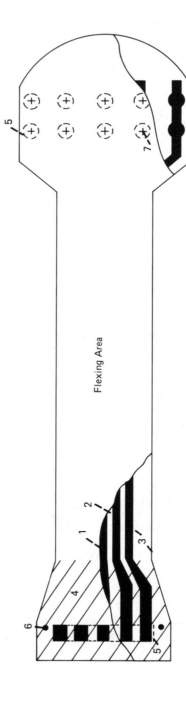

Figure 8.1 General outline of a typical computer-grade circuit. (Courtesy of Rogers Corporation, Chandler, Ariz.)

very closely toleranced, and have specifications precluding any foreign material between the circuit layers, or attached to any outside surfaces. Figure 8.1 shows the general outline of such a typical circuit. Its typical inspection plan might be as follows:

Illustration Points

1. Minimum conductor widths (to be specified by user)
2. Minimum space widths (to be specified by user)
3. Minimum cutline margin widths (to be specified by user)
4. Cross-hatched area indicates adhesive location on back of circuit for mounting
5. Dotted lines indicate cover film openings
6. Hole diameter locations (2 places)
7. Hole diameter locations (10 places)

Electrical

One-hundred-percent test at 100 VDC for minimum insulation resistance of 50 MΩ and continuity of 0.5 Ω maximum.

Visual

One-hundred-percent inspection at magnification of 7.

Conductors

1. Refer to drawing for minimum conductor widths and locations. (See Illustration Point 1.)
2. Isolated imperfections may reduce minimum conductor widths by 50%.

Holes

Inspect for hole diameter, location, sizing, and tolerances. (See Illustration Points 6 and 7.)

Spacing

1. Lines and spaces as small as a 12-mil pitch (6-mil lines and 6-mil spaces) are acceptable.
2. Refer to drawing for minimum space widths and locations. (See Illustration Point 2.)

3. Isolated imperfections may reduce minimum spaces by
    50%.
4. Minor lamination defects at random locations are ac-
    ceptable, providing each delamination or pinhole meets
    the following:

    a. They have no dimension greater than 0.060
        inches.
    b. They do not reduce spacing to less than 50% of
        minimum.
    c. They do not violate minimum cutline margins.
    d. They are no closer than 0.010 inches to access
        holes or the circuit perimeter.

5. Foreign material is acceptable under the following
    conditions:

    a. Particle does not increase circuit thickness more
        than 20%.
    b. Bridging particle is electrically tested and
        accepted.

Solder

1. Dewetted terminations shall have a minimum of 75% cov-
    erage, the remaining 25% may have imperfections such
    as rough spots, pits, etc. Adhesive or film burrs or
    stringers less than 0.010 in. in length are acceptable
    providing all annular ring requirements are met.
2. Solder wicking shall not exceed 0.006 inches. (See
    Illustration Points 6 and 7)
3. Minimum solder annular ring should be 0.003 in. for
    270°. (See Illustration Point 7)
4. Flattened solder pads are acceptable providing pads
    do not lift, delaminate and providing no film slugs are
    left in holes.

Surface and Edge Condition

1. Surface and edges of flexible circuit shall be free from
    visible cuts, chips, bends, loose particles, contami-
    nates, or adhered foreign material.

2. Stringers on edges are acceptable if they do not exceed 0.030 in. in length, or 0.0015 in. in height, width, or diameter.
3. No film shattering will be evidenced in any dynamic area of the circuit.
4. No exposed copper permitted along edges.
5. No edge delamination permitted.
6. Refer to drawing for minimum cutline margin and locations. (See Illustration Point 3.)

Cover Film Alignment

1. All rectangular pads must be captured by coverfilm or adhesive squeezeout. (See Illustration Point 5)
2. Misalignment of coverfilm shall not expose adjacent conductors or pads.

Markings

Each circuit shall be legibly marked with customer part number as required.

Packaging

1. Vendor standard label shall be on the outside of each bag, and shall bear the customer part number, vendor part number, and quantity enclosed.
2. Parts will be clean, without handling damage or foreign material from cleaning operations. Any other obvious defects will be rejected.
3. QA stamp "accepted" label should be affixed to each bag.
4. All bags should be heat sealed.

The preceding testing and inspection plan might be typical for a dynamic computer-grade circuit. Circuits with less critical applications would require inspection plans of much less detail. The main point to remember when designing the circuit, is to balance the function against the cost. An example would be the pointlessness of inspecting a relay interconnect circuit, used in a large industrial switching system, installed in a dirty environment, for minute specs of foreign

material which would have no effect on the functionality of
the circuit.

Testing flexible circuits sometimes deals with more than a
visual or an electrical check. Since there are both static and
dynamic applications, it stands to reason that some testing
for functionality is required to insure that a production circuit will work satisfactorily.

After the circuit is manufactured in either panels or in roll
form, it is cut out, and holes are pierced or drilled where
required. The user then inserts component leads by hand
or with automatic equipment, after which the leads are soldered into place. Only after this is accomplished is the circuit ready for testing. In a static application, since there
is no motion while in use, the testing usually consists of an
electrical screening for function. After passing this initial
screening, the circuit is bent or formed, and put into position. If connectors are a part of the circuit assembly, they
sometimes require a voltage drop test to make certain that
the crimp or mechanical contact is good.

In a dynamic application, the most critical test is the actual
flexing of the circuit to make sure the design allows for an
adequate safety margin of motion cycles before a conductor
failure occurs.

There have been many disappointments among circuit design engineers who design a circuit and believe that it will
serve well for an application when the construction form factor passes a standard industry test such as the IPC test for
flexibility (IPC-TM-650-2.4.3) or (IPC-FC-241A 3.7.6). The
industry standard tests should never be used for product
acceptability unless the method and circuit called out in the
test are *identical* to the circuit in application.

The only way to adequately provide for assurance of flex
quality is to test the production circuit either *in application*,
or in an electrical test fixture (Fig. 8.2) until one or more
of the conductors fail in continuity. The testing required
sometimes takes months and sometimes even a year or more
since some circuits are required to flex over 500 million cycles with no conductor failures! To be absolutely sure of
the circuit life expectancy, a group of circuits from each lot
of every circuit vendor should be qualified and checked to a

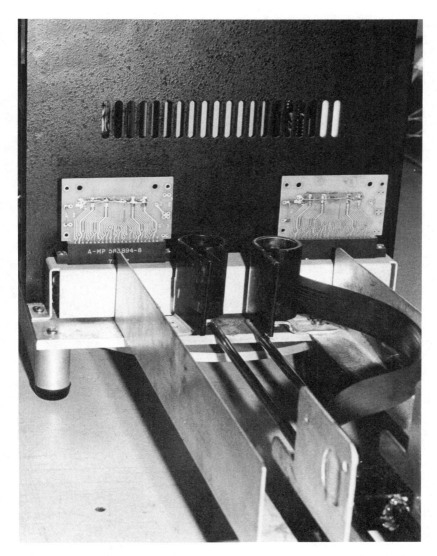

Figure 8.2 To assure flex quality, the production circuit is tested in an electrical test fixture. (Courtesy of Rogers Corporation, Chandler, Ariz.)

desired average quality level (AQL) since flex life varies from manufacturer to manufacturer even when the identical material system is specified!

A proper analysis of all printed wiring applications is necessary *before* the final design is completed so the designer is absolutely sure that material combinations, and circuit trace artwork all work together for the proper form, fit, and function.

9
The Utopian Interconnect

Although many designers would like to think that flexible circuitry is a cureall for their interconnection woes, they will find this is only true if they have analyzed their requirements and compared the analysis with the many options available for a solution. One or more of these solutions *may not* be flexible circuitry!

This chapter will review several applications of flexible circuits which have proven very effective and which have contributed to large cost savings.

9.1 THE INSTRUMENT CLUSTER (FIG. 9.1)

A large number of automobiles have used a flexible circuit to interconnect the instrument cluster of warning lights and gauges for at least 17 years. The cost savings due to ease of installation and the decrease of repairs under warranty have been enormous. In this application, all lamps and gauges used to be interconnected with a confusing wiring harness. This "rats nest" of wires was time consuming in its assembly and installation. A flexible circuit was designed to replace the wiring harness but retain its 3-dimensional design freedom. It was decided to manufacture the flex circuit

Figure 9.1 The instrument cluster. (Courtesy of Sheldahl, Inc., Northfield, Minn.)

from Mylar which was 0.005 in. thick. This would provide a low-cost, but sturdy material which did not have to be exposed to soldering. All connections were designed for mechanically crimped connectors. The copper used was 2 oz, high-elongation, electrodeposited copper which has an excellent surface for bonding the foil to the Mylar with a modified polyester adhesive system. The coverlay, or top insulation layer, was manufactured with 0.003 in. thick Mylar. The adhesive for the cover film was dyed with one of several colors so the circuit would have a "front-side" and "back-side" recognition to the assembly line workers. There have been over 100 million of these circuits manufactured, and in this kind of high-volume production, efficiencies in manufacturing process control have brought the price down to *less* than 1 cent per square inch. This flexible circuit application might be called the utopian circuit because more have been made than any other kind, the application has lasted longer than any other, and its use has saved more money for the user than any other flex circuit.

9.2 THE CONTROL PANEL INTERCONNECT (FIG. 9.2)

This flexible circuit was designed for many of the same reasons the instrument cluster circuits were. The circuit saved
time in assembly, and interconnected potentiometers, switch-
es, and other passive components to the front panel of a res-
pirator. Because all connections were to be soldered, a design had to be developed using high-temperature materials.
The base laminate was made with 0.002-in. thick Kapton polyimide film, to which was bonded 2-oz rolled annealed copper,
with a modified epoxy adhesive system. The cover film was
also 0.002-in. thick Kapton coated with the same modified
epoxy adhesive system. The circuit was designed with
large solder pads. These large pads allowed for the hand-
soldering of the circuit without damage. The size of the pads
acted as copper heat sinks, which helped prevent the pads

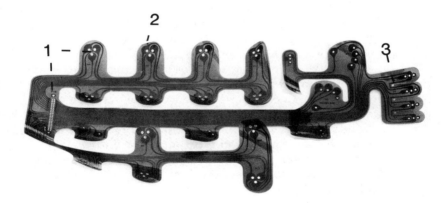

Figure 9.2 The control panel interconnect. (1) Openings in
the cover film which expose solder pads and also cover the
pads 360° to eliminate pad "lift." (2) Large solder pads to
allow the soldering of large components without delamination.
(3) Circular "tear stops" to prevent slits from developing
tears into the circuit. (Courtesy of Rogers Corporation,
Chandler, Arizona.)

from delaminating or lifting from the base insulator. Additionally, the cover film was designed with smaller holes than the pad sizes so each pad was "captured" 360° around the edges. This aided in pad retention during wave-soldering as well. There is a very unusual outline in this circuit which would ordinarily have sharp inside corner radii. These sharp corners are areas which could easily start tears if a twisting stress were applied to the circuit. To help prevent tears in these areas, the corners were given a generous radius and all slits in the material had tear-stop holes punched at the end of each slit. Although this circuit was more expensive than the wires it replaced, it was cost effective in terms of saving space and installation labor.

9.3 A MULTIFUNCTION INTERCONNECT (FIG. 9.3)

This circuit is a combination of a flexible circuit and a printed circuit (PC) hardboard, combined into one interconnect. It is used to connect multiple magnetic disk drive heads to the logic section of a disc memory system. This circuit is constructed in a double-sided, plated-through hole configuration using Kapton as the insulating film, both as the base laminate insulator and the coverlay insulating material on both sides of the circuit. The hardboard, as previously mentioned, is actually a 4-layer multilayer PC hardboard which is bonded to the flexible circuit. When components are mounted on the circuit, the component leads extend through the flex circuit and the hardboard. When the leads are soldered into place, the solder wicks through the entire substructure, joining mechanically and electrically the two sides of the flexible circuit as well as the four layers of the multilayer. This technique of layer interconnection is less costly than a conventional rigid flex circuit process which uses the plated-through-hole process to provide a continuous conductive barrel of copper from the bottom of the 4-layer hardboard up through all of the layers to the top layer of the flexible circuit. Where the flexible circuit extends away from the hardboard, a severe strain can develop if the circuit is flexed back and forth against and away from the board-to-flex intersection. This strain is minimized a great deal by injecting a strip of RTV along the point of interface to

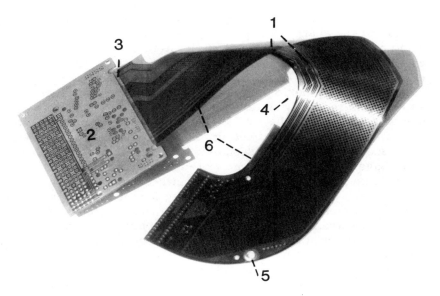

Figure 9.3 A multifunction interconnect. The flexible circuit illustrated above is used by a computer manufacturer to connect multiple magnetic head signals from the head disk assembly and transmit them to the logic section of the disk drive. The circuit illustrated has the following features: (1) A double-sided, plated-through-hole construction using Kapton as the insulating material, diclad with rolled annealed copper, and covered with a Kapton cover film on both sides. (2) A 4-layer multilayer hardboard and a nonclad hardboard attached to opposite ends. (3) A seam of RTV drawn at the point of interface between the hardboards and the flex. This provides a strain relief by forcing a large radius when the circuit is bent and stressed at the points where the hardboards are connected. (4) A formed rubber material is pressed around the inside curved radius to act as an additional strain relief in that area. (5) An eyelet placed in the area of the curved end which serves as a reinforcing agent for a ground lug which is used to ground the circuit to the disk drive. (6) One side of the circuit which serves as conductor routing with varying current carrying capacities. The other side which serves as a grounding bus with integral shield. (Courtesy of Rogers Corporation, Chandler, Ariz.)

provide a soft radius for the flexible circuit to bend against, instead of the sharp edge of the hardboard. An eyelet is located on one end of the circuit and is used to reinforce the circuit when a grounding lug is tightened down on the circuit, which grounds it to the disk-drive chassis.

One advantage flexible circuitry brings to this and other applications requiring it, is the ease of applying a shield to one or both sides of the circuit to prevent radio-frequency (RF) interference from affecting the electronic signals within the traces of the circuit itself. Flexible circuit laminates, in fact, have been used more and more in recent months for groundable RF shields since a low-cost laminate such as 1-oz copper-5-mil polyester material is easily formable, stamps into patterns easily using rule dies, and can be grounded easily to electronic chassis points. The copper portion of the shield material is easily soldered for a good electrical ground.

This particular flexible circuit is no different than others in so far as its main purpose is to save installation costs because of the error-proof wiring, elimination of many connection points, and the saving of time relative to the actual assembly of the circuit into the electronic system it interconnects.

9.4 A MULTIPLANE DYNAMIC INTERCONNECT (FIG. 9.4)

This flexible circuit is used to connect logic output signals to a thermal printhead. The circuit is actually five smaller circuits, which are manufactured separately and then combined in the final manufacturing step to form one homogeneous circuit. It also features a glass epoxy printed circuit hardboard to serve as a support for the five circuit ends as well as a module to serve as a plug which connects to a printed circuit card connector located at the output section of the electronic printer. The tail of each of the five circuits then attach to sections of the thermal printhead.

This kind of circuit must be light weight and be able to withstand many millions of motion cycles with no conductor failures. The individual circuit segments are made by combining 1 mil of Kapton to 1 oz rolled annealed copper and using 1 mil Kapton to insulate the top of the circuit after the

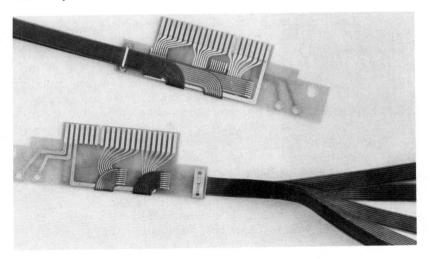

Figure 9.4 A multiplane dynamic interconnect. The features of this product as shown in the illustration are as follows: (1) Lightweight circuit segments (1 mil Kapton and 1-oz rolled, annealed copper × 1 mil Kapton) capable of millions of flex cycles with no conductor failures. (2) Circuit ends folded over to facilitate reflow solder operation. (3) A G-10 polarized hardboard to which the flexible circuits are connected with reflowed solder to provide a low profile and eliminate expensive connectors. (4) A strain relief staple to eliminate mechanical strain at the soldered joints. (5) Goldplated conductors to insure a low contact resistance mating with the system edgecard connector. (Courtesy of Rogers Corporation, Chandler, Ariz.)

traces are etched. This puts the copper conductors in a neutral axis so there will not be unequal stress on the copper during the operation of the system.

The individual circuit ends are folded over to facilitate soldering to the printed circuit hardboard traces. This eliminates the need to design and manufacture a circuit with a double-sided access feature, which is much more costly than a standard single-sided circuit construction. The glass epoxy hardboard circuit is notched in one area between its plug traces. This provides polarization os the circuit module cannot be plugged in the wrong way.

Because this is a moving circuit, there is strain which is exerted at each soldered connection. A staple is used at the end of the hardboard, which ties down all of the five individual circuits, one on top of the other, to provide a strain relief. The traces on the hardboard are also goldplated to provide very low contact resistance. These traces mate with the card edge connector's goldplated contacts. This type of circuit is sometimes called a hybrid because of the combination of flexible circuit and hardboard printed circuit.

It is apparent by reviewing this discussion that flexible circuits are as varied as a designer's dream can wander. It is also very clear that flexible circuit vendors are truly job shops. Because this is true, it is imperative that the company thinking of using flexible circuitry first lay out a preliminary design and then *visit* prospective vendors so that one is familiar with the processes and capabilities of each.

This business is very competitive and vertically oriented so one company may be better than another, depending on the type of circuit desired. Some companies specialize in military products, some in consumer products, while others are more proficient in computer or telecommunications products.

Care and planning will translate into time well spent when it comes to vendor selection and the finalization of the design. A well-designed flexible circuit will lead to a successful interconnect device which will improve efficiency and reduce cost.

10
Overview

This overview deals with applications analyses and the selection of vendors. Although this subject may be somewhat controversial, the fact remains that all flexible circuits can be grouped into classes, and most flexible circuit vendors can be grouped into classes as they relate to expertise in manufacturing a certain class of flexible circuits. In short, there is no flexible circuit manufacturer who is all things to all people!

Flexible circuitry can be generally grouped into categories. These categories can be thought of as markets, and broken down accordingly. The Institute for Interconnection and Packaging Electronic Circuits (IPC) listed seven market types in their 1982 report by the Technology/Marketing Research Council. These market types, along with the market share of each is listed in the following chart (Table 10.1). This chart also shows the estimate of free-market producers of flexible circuits along with the production value of captive producers, or companies producing circuits for internal consumption.

Table 10.1 1982 Market Value of All Types of Flexible Circuits Includes Data on Independents and Estimates For Captive Producers (All Dollars in Thousands – 000 Omitted)

	Value of market independent producers of flex circuits		Estimated value of captive production		Total market	
	Dollars	% of Total	Dollars	% of Total	Dollars	% of Total
Business retail	1,326	1.3	1,000	1.2	2,326	1.3
Communications	4,896	4.8	40,000	50.0	44,896	24.7
Consumer	13,974	13.7	5,000	6.3	18,974	10.4
Computer	29,274	28.7	10,000	12.5	39,274	21.6
Gov't/military	41,310	40.5	20,000	25.0	61,310	33.7
Industrial	7,548	7.4	2,000	2.5	9,548	5.2
Instrumentation	3,672	3.6	2,000	2.5	5,672	3.1
Totals	102,000	100%	80,000	100%	182,000	100%

Table 10.2 Type of Material Used in the Manufacture of Flexible Circuits

Type of material	Value in $
Polyester	7 million
Nomex	< 1 million
Polyetherimide	< 1 million
Reinforced composite	6 million
Polyimide	16 million

Some experts break the market down into types based on the kind of material or substrate used in the manufacture of flexible circuits. This type of market breakdown can be shown as follows (Table 10.2). Remember that the cost of various substrates differs markedly, and the dollars of value should not be construed the same as area of material produced.

Although most flexible circuits can be categorized by base material type—polyester for automotive and communications circuits and polyimide for military and computer circuits, it's not enough to use this simplistic approach when deciding into which category your particular design falls or the list of vendors best equipped to handle the manufacture of your circuit. The following information may help with your decision. The major independent manufacturers of flexible circuits, in alphabetical order are:

AT&T Technologies, Inc.	Parlex
Cirtel	Rogers Corporation
Flexible Circuits Inc.	Sanders Associates
Hughes Aircraft	Sheldahl Inc.
Interconics	Teledyne Electro Mechanisms

Other manufacturers of small volumes of flexible circuits are:

Applied Circuit Technology	Electro Film Inc.
Advanced Flex, Inc.	G.C. Aero
AMP Incorporated	General Circuits
Cal Flex	Graphics
Century Circuits	Methode
Com Cir Tech	Minco
Dytronics Division, GTI Electronics	Tektronix, Inc., Applied Chemical Components Group
	Universal Circuits

Obviously if you contacted all of the above companies you could get a huge range of responses from no bids to very high prices. It could be best then to categorize circuit types and match these types to specific manufacturers.

Circuit category	Characteristics of design	Cost to produce
High-volume consumer (automobiles and communications equipment)	Wide lines and spaces, loose tolerances, low-cost materials (e.g.: polyester, ED copper)	Low
Nonaerospace military ground systems and fuze interconnects	Medium to fine lines and spaces, high-temperature requirements for polyimide materials, acrylic or polyimide adhesives	Medium to high
Computer	Medium to fine lines and spaces, high dynamic requirements close tolerances	Medium to high
Instrumentation	Medium lines and spaces, may be static or dynamic in application, demanding but not tight tolerances	Medium
Aerospace	Medium to fine lines and spaces, high temperature requirements, high information density requiring many layers, sometimes with rigidizing members, use polyimide films, acrylic or polyimide adhesives.	High

Using these descriptive categories, the following list of companies could be considered as vendors.

Company	Location code*	Circuit type
AT & T Technologies, Inc, Marketing Services Dept. 110, Allentown, PA	215	1,2,4
Cirtel, Irvine, Calif.	603	3,4
Flexible Circuits Inc., Warrington, Pa.	714	5
Hughes Aircraft, Irvine, Calif.	215	2,4,5
Interconics, Nashua, N.H.	714	5
Parlex, Methuen, Mass.	617	2,4,5
Rogers Corporation, Chandler, Ariz.	602	2,3,4
Sanders Associates, Nashua, N.H.	603	5
Sheldahl Inc., Northfield, Minn.	507	1,2,4
Teledyne Electro Mechanisms, Hudson, N.H.	603	2,5
Tektronix, Inc., Applied Chemical Components Group, Beaverton, Ore.	503	3,4

The following companies manufacture circuits in smaller volumes and can sometimes provide good prototype services with fast turnaround.

Company	Location code	Circuit type
Advanced Circuit Technology, Nashua, N.H.	603	2,4
Advanced Flex Inc., Minneapolis, Minn.	612	2,3,4,5
AMP Inc., Salem, N.C.	919	3,4
Cal Flex, Anaheim, Calif.	714	2,3,4

---

*The location code is the telephone area code for each manufacturer. They can be contacted by calling telephone company information for their telephone number.

Company	Location code	Circuit type
Century Circuits, St. Paul, Minn.	612	2,3,4,5
Com Cir Tech Inc., Buffalo, N.Y.	716	4,5
Dytronics Div. GTI Elec- tronics, Leesberg, Ind.	219	1,2,4
Electro Film Inc., Saugus, Calif.	805	3,4,5
G.C. Aero, Torrance, Calif.	213	4

Using this guide, selection of several proper vendors to produce your circuit should be relatively easy. Remember, a personal vendor review is highly recommended before a final selection is made. This will give you extra confidence in your selection of the right flexible circuit manufacturer.

11
Conclusion

Even though this book covers most of the basic knowledge
needed for initial flexible circuit design, this industry, like
others of high technology, changes very rapidly with advan-
ces in knowledge and with engineering requirements for hardware interconnects. For this reason it is desirable to belong
to a trade organization which promotes business through technical publications, seminars, and trade meetings. These activities promote the growth objectives of the industry and are
one, if not *the* most important link between users and suppliers of products.

The major trade organization supporting the printed circuit
industry is The Institute for Interconnecting and Packaging
Electronic Circuits (IPC). The headquarters for this organization is located at 3451 Church Street, Evanston, Illionois
60203.

Because of a need for technical information, and to
develop standards in the field of printed wiring,
the IPC was formed in 1957. . . The programs of
the IPC are made possible only through the active,
voluntary support of the membership. The membership of the IPC is made up of representatives of

companies from a broad cross-section of industry,
and also includes qualified technical experts from
Government Agencies, and representatives from
colleges and universities. Membership is on a company basis and members include companies that
produce printed wiring boards (including flexible
circuits) for sale; companies that produce printed
wiring boards for internal use, . . . and companies
which are suppliers of material and equipment used
in the industry.

The IPC makes available to members and nonmembers alike,
a proliferation of technical resource information including
video tapes, over 35 specifications and standards, design
guides, as well as technical reports. Committees are organ-
ized by specific interests and are attended by all members
who wish to participate in the special interests represented
by their companies. This type of organization provides a
group of industry experts, which for synergistic reasons,
create a forum of up-to-date, outstanding technical
excellence.

Section six of the *IPC Design Guide*, volume 1 contains information on the design of flexible circuits. It covers single-
and double-sided, as well as multilayer design guidelines.
After reading section two on "General Printed Circuit Information" the section on flexible circuits covers the subject
very well. It was developed from data taken from the work
of over 30 technical committees of the IPC as well as from
other recognized sources of technical data.

One of the most important things to remember about flexible
circuit design, is that there are different manufacturing technologies used, depending on the products and tolerances
desired. Let me point out two examples of what I mean.

If a manufacturer has a requirement for a flexible circuit
which is used to interconnect an automobile instrument cluster such as the product shown in the previous chapter, that
circuit could be described as a device to interconnect an area
of over $1/2$ ft^2. It must be low in cost. It will have broad
conductor traces and spaces, and its tolerances can be relatively loose. The manufacturing facility that produces this
product should be relatively clean, the image resist will be a
low-cost screened product, the material will be made up of

polyester film for the base and cover film. It will be processed entirely by the roll-to-roll process. Because of relatively loose tolerances, roll-to-roll visual inspection will be performed over a light-table. This kind of circuit can be produced in production quantities for less than 1 cent per square inch.

The second example is a circuit which interconnects a flying magnetic head on a computer disk-drive assembly to its read/write logic. This circuit is small, probably 10-15 in^2. It is farily dense with conductor traces and spaces as small as 0.006 in. The tolerances relative to artwork with respect to tooling and cover film registration are tight, ($\pm$0.003). The circuit must withstand high temperatures of soldering, and it must withstand many flexing operations with no conductor failures (as many as 500 million cycles). The company that produces this circuit will have most of its operations in a class 10,000 clean-room environment. The circuit image will be produced with a photoresist using dry film, the circuit construction will be a base of polyimide film with a polyimide cover film. The copper will be specially selected and treated rolled annealed foil. Close tolerances will require inspection of each circuit under magnification. Because of the extra cost components of this product, it will be produced in production volumes for as much as 25 cents per square inch.

As you can see, it requires not only a different kind of design attention, production facility, and inspection criteria, but a totally different production "mind-set" to be effective in producing the part efficiently.

In conclusion, once you have decided to use a flexible circuit for your interconnect, make sure its design and manufacturing requirements match the capabilities of the flexible circuit manufacturer. Things to look for during a vendor survey, are roll-to-roll processing clean-room capabilities with precisely controlled temperature and humidity, semiautomatic screening, semiautomatic and optical control of indexing raw materials, good hard tooling capability, and good artwork-producing capabilities, along with roll-to-roll dry film imaging capabilities.

As the flexible circuitry technology changes and improves, new processes and procedures will be used, such as fully additive, roll-to-roll, vapor deposition of circuits, semiadditive processes using much thinner electrolytically deposited copper on film to insure felxibility, and which can also be etched at rates much faster than is presently done with heavier copper. Also, we will see laser-drilled holes, only 0.002 in. in diameter which will allow circuits dense enough to carry three times as many lines of information and allow for direct bonding of semiconductor devices. The possibilities for flexible circuit interconnects will be limited only by the imagination of the design engineer!

Glossary

Activating. A treatment that renders nonconductive material receptive to ejectroless deposition. (Nonpreferred synonyms: Seeding, Catalyzing, and Sensitizing.)

Additive process. A process for obtaining conductive patterns by the selective deposition of conductive material on unclad base material. (See also **Semiadditive process and Fully-additive process.**)

Adhesion promotion. The chemical process of preparing a plastic surface to provide for a uniform, well-bonded metallic overplate.

Ambient. The surrounding environment coming into contact with the system or component in question.

This glossary is an excerpt from ANSI/IPC-T-50B and is used with permission of the Institute for Interconnections and Packaging Electronic Circuits, Evanston, Ill.

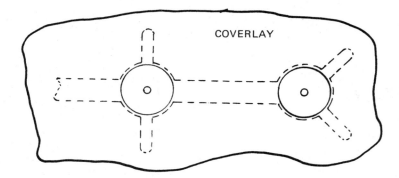

Figure 1. Lands with anchoring spurs.

Anchoring spurs. Extnesions of the lands on flexible printed wiring, extending beneath the cover layer to assist in holding the lands to the base material. (See Fig. 1.)

Annular ring. That portion of conductive material completely surrounding a hole.

Artwork. An accurately scaled configuration which is used to produce the artwork master or production master (See Fig. 2.)

Artwork master. An accurately scaled (usually 1:1) pattern which is used to produce the production master. (See Fig. 2.)

Assembly. A number of parts or subassemblies or any combination thereof joined together to perform a specific function.

Note: When this term is used in conjunction with other terms listed herein, the following definitions shall prevail.

- **Printed wiring assembly.** A printed wiring board on which separately manufactured components and parts have been added.

- **Multilayer printed wiring assembly.** A multilayer printed wiring board on which separately manufactured components and parts have been added.

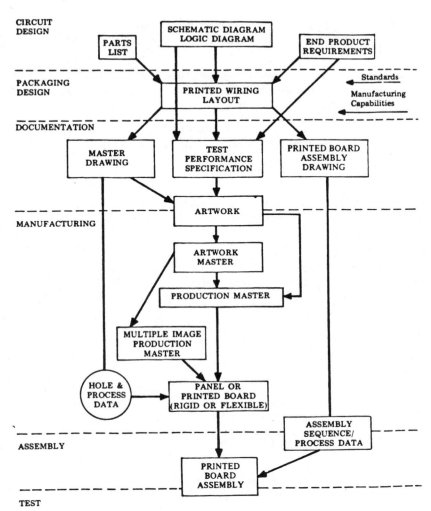

Figure 2.

- **Printed circuit assembly.** A printed circuit board on which separately manufactured components and parts have been added.

- **Multilayer printed circuit assembly.** A multilayer printed circuit board on which separately manufactured components and parts have been added.

- **Printed board assembly.** An assembly of several printed circuit assemblies or printed wiring assemblies.

B

Base material. The insulating material upon which the conductor pattern may be formed. The base material may be rigid or flexible. It may be a dielectric sheet or insulated metal sheet.

Base material thickness. The thickness of the base material excluding metal foil or material deposited on the surfaces.

Basic dimension. A numerical value used to describe the theoretical exact location of a feature or hole. It is the basis from which permissible variations are established by tolerances on other dimensions, in notes, or by feature control symbols.

Blank. An unprocessed or partially processed piece of base material, or metal-clad base material, cut from a sheet or panel and having the rough dimensions of a printed board.

Bleeding. A condition in which a plated hole discharges process material or solution from crevices or voids.

Blister. A localized swelling and separation between any of the layers of a laminated base material, or between base material and conductive foil. (It is a form of delamination.)

Board thickness. The thickness of the metal-clad base material including conductive layer or layers. (May include additional platings and coatings depending upon when the measurement is made.)

Bond strength. The force per unit area required to separate two adjacent layers of a board by a force perpendicular to the board surface. (See also **Peel strength.**)

Bridging electrical. The formation of a conductive path between conductors.

B-staged resin. A resin in an intermediate state of cure. (The cure is normally completed during the laminating cycle.)

C

Capacitive coupling. The electrical interaction between two conductors caused by the capacitance between them.

Catalyzing.* See **Activating**.

Center-to-center spacing. The nominal distance between the centers of adjacent features on any single layer of a printed board. (See also **Pitch**.)

Certification. Verification that specified testing has been performed, and required parameter values have been attained.

Circumferential separation. A crack (1) in the plating extending around the entire circumference of a plated-through hole, or (2) in the solder fillet around the lead wire, or (3) in the solder fillet around an eyelet, or (4) at the interface between a solder fillet and a land.

Chamfer. A broken corner to eliminate an otherwise sharp edge.

Characteristic impedance. The ratio of voltage ot current in a propagating wave, i.e., the impedance which is offered to this wave at any point of the line. (In printed wiring its value depends on the width of the conductor, the distance from the conductor to ground plane(s), and the dielectric constant of the media between them.)

Chemical hole cleaning. The chemical process for cleaning conductive surfaces exposed within a hole. (See also **Etchback**.)

---

*The preferred term is referenced.

Chemically-deposited printed wiring or **Chemically-deposited printed circuit.*** See Additive Process.

Clad. (Adj.). A condition of the base material to which a relatively thin layer or a sheet of metal foil has been bonded to one or both of its sides, e.g., a metal-clad base material.

Clearance hole. A hole in the conductive pattern larger than, but coaxial with, a hole in the printed board base material. (See Fig. 3.)

Clinched leads. Component leads which extend through the printed board and are formed to effect a spring action, metal-to-metal electrical contact wtih the conductive pattern prior to soldering. (See Fig. 4.)

Clinched-wire interfacial connection.* See Clinched-wire through connection.

Clinched-wire through connection. A connection made by a wire which is passed through a hole in a printed board, and subsequently formed, or clinched, in contact with the conductive pattern on each side of the board, and soldered. (See Fig. 5.)

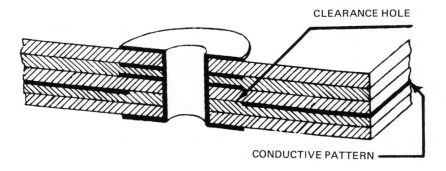

CLEARANCE HOLE

CONDUCTIVE PATTERN

Figure 3. Clearance hole.

---

* The preferred term is referenced.

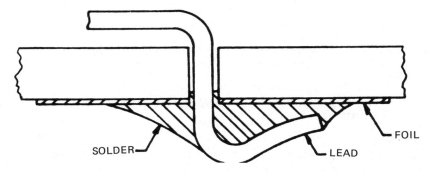

Figure 4. Clinched lead termination (after soldering).

Cold solder joint. A solder connection exhibiting poor wetting and a grayish, porous appearance due to insufficient heat, inadequate cleaning prior to soldering, or to excessive impurities in the solder solution.

Component density. The quantity of components on a printed board per unit area.

Component lead. The solid or stranded wire or formed conductor that extends from a component and serves as a mechanical or electrical connection or both.

Component hole. A hole used for the attachment and electrical connection of component terminations, including pins and wires, to the printed board.

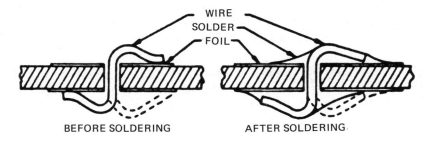

Figure 5. Clinched-wire through connection.

Component side. That side of the printed board on which most of the components will be mounted.

Conditioning. Time-limited exposure of a test specimen to a specified environment(s) prior to testing.

Conductive foil. A thin sheet of metal that may cover one or both sides of the base material and is intended for forming the conductive pattern.

Conductive pattern. The configuration or design of the conductive material on the base material. (Includes conductors, lands, and through connections when these connections are an integral part of the manufacturing process.)

Conductor. A single conductive path in a conductive pattern.

Conductor base width. The conductor width at the plane of the surface of the base material. (See also **Conductor width** and **Design width of conductor.**)

Conductor layer. The total conductive pattern formed upon one side of a single layer of base material.

Conductor layer no. 1. The first layer having a conductive pattern, or a multilayer printed board, on or adjacent to the component side.

Conductor pattern.* See Conductive Pattern.

Conductor side. The side of a single-sided printed board containing the conductive pattern.

Conductor spacing. The distance between adjacent edges (not centerline to centerline) of isolated conductive patterns in a conductor layer.

Conductor thickness. The thickness of the conductor including all metallic coatings. (It excludes nonconductive protective coating.)

Conductor width. The observable width of a conductor at any point chosen at random on the printed board, normally viewed from vertically above unless otherwise specified.

---

*The preferred term is referenced.

(Imperfections, for example nicks, pinholes, or scratches, allowable by the relevant specification shall be ignored.) (See also **Design width of conductor** and **Conductor base width.**)

Conductor to hole spacing. The distance between the edge of a conductor and the edge of a supported or unsupported hole.

Conformal coating. An insulating protective coating, which conforms to the configuration of the object coated, applied to the completed board assembly.

Connector area. That portion of printed wiring used for the purpose of providing external electrical connections.

Contact area. The common area between a conductor and a connector through which the flow of electricity takes place.

Contact spacing. The distance between the centerlines of adjacent contact areas.

Continuity. An uninterrupted path for the flow of electrical current in a circuit.

Corner mark (crop mark). The mark at the corners of a printed board artwork, the inside edges of which usually locate the borders and establish the contour of the board.

Coupon.* See **Test Coupon.**

Cover lay, cover layer, cover coat. An outer layer(s) of insulating material applied over the conductive pattern on the surface of the printed board.

Cracking. A condition consisting of breaks in metallic or nonmetallic coatings or both that extend through to an underlying surface.

Crazing. An internal condition occurring in the laminated base material in which the glass fibers are separated from the resin at the weave intersections. This condition manifests itself in the form of *connected* white spots or "crosses" below

---

*The preferred term is referenced.

the surface and the base material, and is usually related to *mechanically* induced stress.

Crazing (conformal coating). A network of fine cracks on the surface or within the conformal coating.

Crosshatching. The breaking of large conductive areas by the use of a pattern of voids in the conductive material.

Crosstalk. The undesirable interface caused by the coupling of energy between signal paths.

Current-carrying capacity. The maximum current which can be carried continuously, under specified conditions, by a conductor without causing objectionable degradation of electrical or mechanical properties of the printed board.

D

Datum reference. A defined point, line, or plane used to locate the pattern or layer for manufacturing, inspection, or for both purposes.

Delamination. A separation between plies within the base material, or between the base material and the conductive foil, or both.

Dent. A smooth depression in the conductive foil which does not significantly decrease foil thickness.

Design width of conductor. The width of a conductor as delineated or noted on the master drawing. (See also **Conductor base-width**, **Conductor width**, and **Current-carrying capacity**.)

Device. An individual electrical element, usually in an independent body, which cannot be further reduced without destroying its stated function.

Dewetting. A condition which results when molten solder has coated a surface and then receded leaving irregularly shaped mounds of solder separated by areas covered with a thin solder film; base metal is not exposed.

Dielectric breakdown. A complete failure of a dielectric material characterized by a disruptive electrical discharge through the material due to a sudden and large increase in voltage.

Dielectric constant. The ratio of the capacitance (C_X) of a given configuration of electrodes with a specified dielectric, to the capacitance (C_V) of the same electrode configuration with a vacuum (or air) as the dielectric.

Dielectric strength. The maximum voltage that a dielectric can withstand, under specified conditions, without resulting in a voltage breakdown (usually expressed as volts/unit dimension).

Digitizing. Any method of reducing feature locations on a flat plane to give digital representation of X-Y coordinates.

Dimensional stability. A measure of dimensional change caused by such factors as temperature, humidity, chemical treatment, age, or stress (usually expressed as Δunits/ unit).

Dimensioned hole. A hole in a printed board where the means of determining location is by coordinate values not necessarily coinciding with the stated grid.

Dip soldering. A process whereby printed boards with attached components are brought in contact with the surface of a static pool of molten solder for the purpose of soldering the entire exposed conductive pattern and component leads in one operation.

Dissipation factor. A measure of the A.C. loss. The Dissipation Factor is proportional to the power loss per cycle (f) per potential gradient (E) (squared) per unit volume as follows:

$$\text{Diss. fac.} = \frac{\text{Power loss}}{E^2 \times f \times \text{volume} \times \text{constant}}$$

Double-sided board. A printed board with a conductive pattern on both sides.

Drag soldering. A process whereby supported, moving printed circuit assemblies or printed wiring assemblies are brought in contact with the surface of a static pool or molten solder.

Dual in-line package (DIP). A component which terminates in two straight rows of pins or lead wires.

E

Edge-board connector. A connector designed specifically for making removable and reliable interconnections between the edge board contacts on the edge of a printed board and external wiring.

Edge-board contacts. A series of contacts printed on or near any edge of a printed board and intended for mating with an edge-board connector.

Edge definition. The fidelity of reproduction of a pattern edge relative to the production master.

Edge spacing. The distance of a pattern, components, or both, from the edges of the printed board. (See also Margin.)

Electroless deposition. The deposition of metal from an auto-catalytic plating solution without application of electrical current.

Electrodeposition. The deposition of a conductive material from a plating solution with application of electrical current. (Nonpreferred synonyms: Electrolytic deposition, and Galvanic deposition.)

Epoxy smear.[*] See Resin Smear.

Etchant. A solution used to remove, by chemical reaction, the unwanted portion of material from a printed board.

Etchback. A process for the controlled removal of nonmetallic materials from sidewalls of holes to a specified depth. It is used to remove resin smear and to expose additionan internal conductor surfaces.

Etch factor. The ratio of the depth of etch (conductor thickness) to the amount of lateral etch (undercut).

----

[*]The preferred term is referenced.

Etched printed board. A board having a conductive pattern formed by the chemical removal of unwanted portions of the conductive foil.

Etching. A process wherein a printed pattern is formed by chemical, or chemical and electrolytic removal of the unwanted portion of conductive material bonded to a base.

Etching Indicator. A wedge-shaped or other specified pattern affixed to the conductive foil to indicate the quality of etching. (See Fig. 6.)

Extraneous copper. Unwanted copper remaining on the base material after chemical processing.

Eyelet. A hollow tube inserted in a terminal or printed board to provide mechanical support for component leads or electrical connection.

F

Fingers.* See Edge-board contacts.

First article. A part or assembly manufactured prior to the start of production for the purpose of assuring that the manufacturer is capable of manufacturing a product which will meet the requirements.

Flat cable. A cable with two or more parallel, round or flat conductors in the same plane encapsulated by an insulating material.

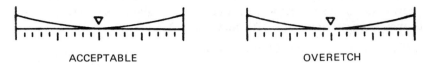

ACCEPTABLE   OVERETCH

Figure 6 Etching indicator.

---

*The preferred term is referenced.

Flexible printed circuit. A random arrangement of printed circuit and components utilizing flexible base materials with or without flexible cover layers.

Flexible printed wiring. A random arrangement of printed wiring utilizing flexible base material with or without flexible cover layers.

Flexural failure. A conductor failure caused by repeated flexing. It is indicated by an increase of resistance measured for a specified time.

Flow Soldering.* See Wave soldering.

Fully-additive process. (Nonpreferred synonym: Fully electroless). An additive process wherein the entire thickness of electrically isolated conductors is built up by electroless metal deposition.

Fused coating. A metallic coating (usually tin or solder alloy) which has been melted and solidified forming a metallurgical bond to the basis metal.

Fusing. The melting of a metallic coating (usually electrodeposited), followed by solidification.

G

Glass transition temperature. The temperature at which an amorphous polymer (or the amorphous regions in a partially crystalline polymer) changes from a hard and relatively brittle condition to a viscous or rubbery condition. (This transition generally occurs over a relatively narrow temperature region, many physical properties undergo significant rapid changes. Some of those properties are hardness, brittleness, thermal expansion, specific heat, etc.)

Graded wedge.* See Etching indicator.

Grid. An orthogonal network of two sets of parallel equidistant lines used for locating points on a printed board.

_____

*The preferred term is referenced.

Ground plane. A conductor layer, or portion of a conductor layer (usually a continuous sheet of metal with suitable ground plane clearances), used as a common reference point for circuit returns, shielding, or heat sinking.

H

Haloing. Mechanically induced fracturing or delaminating on or below the surface of the base material; it is usually exhibited by a light area around holes, other machined areas, or both.

Heat sinking plane. A continuous sheet of metal on or in a printed board that functions to dissipate heat away from heat sensitive components.

Hole breakout. A condition in which a hole is not completely surrounded by the land. (See Fig. 7.)

Hole density. The quantity of holes in a printed board per unit area.

Hole location. The dimensional location of the center of a hole.

Hole pattern. The arrangement of all holes in a printed board.

Hole pull strength. The force necessary to rupture a plated-through hole when loaded or pulled in the direction of the axis of the hole.

Hole void. A void in the metallic deposit of a plated-through hole exposing the base material.

Figure 7. Hole breakout.

I

Icicle.* See Solder projection.

Immersion plating (Galvanic displacement). The chemical deposition of a thin metallic coating over certain basis metals by a partial displacement of the basis metal.

Inclusion. A foreign particle, metallic or nonmetallic, in a conductive layer, plating, or base material.

Indentation.* See Dent.

Index edge, index edge marker, indexing hole, indexing notch, indexing slot.* See Tooling feature.

Initiating.* See Activating.

Inspection lot. A collection of units of product bearing identification and treated as a unique entity from which a sample is to be drawn and inspected to determine conformance with the acceptability criteria.

Inspection overlay. A positive or negative transparency made from the production master and used as an inspection aid.

Insulation resistance. The electrical resistance of the insulating material (determined under specified conditions) between any pair of contacts, conductors, or grounding devices in various combinations.

Ionizable contaminants. Process residues such as flux activators, finger prints, etching and plating salts, etc., that exist as ions and when dissolved increase electrical conductivity.

J

Jumper. An electrical connection between two points on a printed board added after the intended conductive pattern is formed.

---

*The preferred term is referenced.

Jumper wire. A wire used as a jumper.

K

Key. A device designed to assure that the coupling of two components can occur in only one position.

Keying slot. A slot in a printed board which perm ts the printed board to be plugged into its mating receptacle but prevents it from being plugged into any other receptacle. (See also Polarizing slot.)

Keyway. A general term covering both keying slot(s) and polarizing slot(s).

L

Laminate (noun). A product made by bonding together two or more layers of material.

Laminate thickness. Thickness of the metal-clad base material, single or double-sided, prior to any subsequent processing. (See also **Board thickness.**)

Land. A portion of a conductive pattern usually, but not exclusively, used for the connection, or attachment, or both of components.

Landless hole. A plated-through hole without a land(s).

Lead mounting hole.* See Component Hole.

Lead projection. The distance which a component lead protrudes through the printed board on the side opposite from which the component is mounted.

Legend. A format of letters, numbers, symbols, and patterns on the printed board primarily used to identify component locations and orientation, for convenience in assembly and replacement operations. (See also Marking.)

Line.* See Conductor.

---

*The preferred term is referenced.

Locating edge, Locating edge marker, Locating hole, Locating notch, Locating slot.* See Tooling Feature.

M

Margin. The distance between the reference edge of a flat cable and the nearest edge of the first conductor. (See also Edge spacing.)

Marking. A method of identifying printed boards with part number, revision letter, manufacturers code, etc. (See also Legend.)

Mask.* See Resist.

Master dot pattern.* See Hole Pattern.

Master drawing. A document that shows the dimensional limits or grid locations applicable to any or all parts of a printed board (rigid or flexible), including the arrangement of conductive and nonconductive patterns or elements; size, type, and location of holes; and any other information necessary to describe the product to be fabricated. (See Fig. 2.)

Master line.* See Design width of conductor.

Master pattern.* See Production master.

Metal-clad base material. Base material covered with foil on one or both of its sides.

Metal-clad laminate.* See Metal-clad base material.

Metallization. A deposited or plated thin metallic film used for its protective or electrical properties.

Microstrip. A type of transmission line configuration which consists of a conductor over a parallel ground plane, and separated by a dielectric.

Microsectioning. The preparation of a sepcimen for the microscopic examination of the material to be examined (usually by cutting out a cross-section, followed by encapsulation, polishing, etching, staining, etc.)

---

*The preferred term is referenced.

Minimum annular ring. The minimum width of metal, at the narrowest point, between the edge of the hole and the outer edge of the land. This measurement is made to the drilled hole on internal layers of multilayer printed boards and to the edge of the plating on outside layers of multilayer boards and double sided boards.

Minimum electrical spacing. The minimum allowable distance between adjacent conductors that is sufficient to prevent dielectric breakdown or corona or both between the conductors at any given voltage and altitude.

Misregistration. The lack of dimensional conformity between successively produced features or patterns. (See **Registration.**)

Mounting hole. A hole used for the mechanical mounting of a printed board or for the mechanical attachment of components to the printed board.

Multiple-image production master.* See **Production master.**

N

Negative (noun). An artwork, artwork master, or production master in which the intended conductive pattern is transparent to light, and the areas to be free from conductive material are opaque.

Negative etchback. Etchback in which inner conductor layer material is recessed relative to the surrounding base material (See Fig. 8.)

Negative-acting resist. A resist which is polymerized (hardened) by light and which, after exposure and development remains on the surface of a laminate in those areas which were under the transparent parts of a production master.

Nonconductive pattern. A configuration formed by functional nonconductive material of a printed circuit (e.g., dielectric, resist, etc.)

---

*The preferred term is referenced.

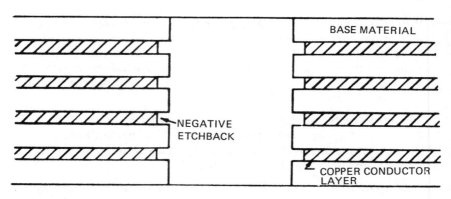

Figure 8. Negative etchback.

Nonfunctional land. A land on internal or external layers, not connected to the conductive pattern on its layer.

Nonpolar solvents. Solvents which are not ionized sufficiently to be electrically conductive and which cannot dissolve polar compounds such as inorganic salts, but can dissolve nonpolar compounds, such as hydrocarbons and resins.

Nonwetting. A condition whereby a surface has contacted molten solder, but the solder has not adhered to all of the surface; base metal remains exposed.

O

Offset land. A land which intentionally not in physical contact with its associated component hole.

One-sided board.* See **Single-sided board.**

Opaquer. A material that, when added to the resin system, renders a laminate sufficiently opaque so that the yarn or weave of the reinforcing material cannot be seen with the unaided eye, using either reflected or transmitted light.

---

*The preferred term is referenced.

Outgassing. Deaeration or other gaseous emission from a printed board assembly (printed board, component or connector) when exposed to a reduced pressure, or heat, or both.

Overhang. The sum or outgrowth and undercut. (See Fig. 10.) (If undercut does not occur, the overhang is the outgrowth only.)

Oxide transfer.* See Treatment transfer.

P

Packaging density. Quantity of functions (components, interconnection devices, mechanical devices) per unit volume, usually expressed in qualitative terms, such as high, medium, or low.

Pad.* See Land.

Panel. A rectangular or square base material of predetermined size intended for or containing one or more printed boards and, when required, one or more test coupons.

Panel plating. The plating of the entire surface of a panel (including holes.)

Pattern. The configuration of conductive and nonconductive materials on a panel or printed board. (Pattern denotes also the circuit configuration on related tools, drawings, and masters.)

Pattern plating. The selective plating of a conductive pattern.

Peel strength. The force per unit width required to peel the conductor or foil from the base material.

Photographic reduction dimension. Dimensions (e.g., the distance between lines or between two specified points) on the artwork master to indicate to the photographer the extent to which the artwork master is to be photographically reduced. (The value of the dimension refers to the 1:1 scale and must be specified.)

---

*The preferred term is referenced.

Photomaster.* See Artwork master.

Pilot hole.* See Locating hole.

Pinhole. A small hole occurring as an imperfection which penetrates entirely through a layer of material.

Pit. A depression in the conductive layer that does not penetrate entirely through it.

Pitch. The nominal distance from center-to-center of adjacent conductors. (Where conductors are of equal size, and spacing is uniform, the pitch is usually measured from the reference edge of a conductor to the referenced edge of the adjacent conductor.) (See Fig. 9.)

Plate finish (pertaining to laminating). The finish present on the metallic surface of metal-clad base material resulting from direct contact with the laminating press plates without modification by any subsequent finishing process.

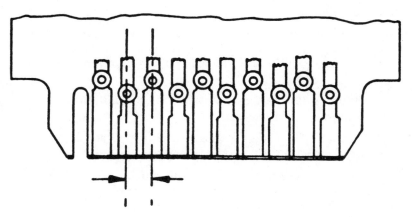

Figure 9. Pitch.

---

*The preferred term is referenced.

Plated-through hole. A hole in which electrical connection is made between internal or external conductive patterns, or both, by the deposition of metal on the wall of the hole.

Plated-through hole structure test. A visual examination of the metallic conductors and plated-through holes of a printed board after the glass-plastic laminate has been dissolved away.

Plating bar. The temporary conductive path interconnecting areas of a printed board to be electroplated, usually located on the panel outside of the borders of such a board.

Plating up. The process consisting of the electro-chemical deposition of a conductive material on the base material (surface holes, etc.) after the base material has been made conductive.

Plotting. The practice of mechanically converting X-Y positional information into a visual pattern, such as artwork.

Polarization. A technique of eliminating symmetry within a plane so that parts can be engaged in only one way in order to minimize the possibility of electrical and mechanical damage or malfunction.

Polarizing slot. A slot, at the edge of a printed board, used to assure proper insertion and location in a mating connector. (See also Keying slot.)

Polar solvents. Solvents which are ionized sufficiently to be electrically conductive and which can dissolve polar compounds, such as inorganic salts, but cannot dissolve non-polar compounds, such as hydrocarbons and resins.

Positive (noun). An artwork, artwork master, or production master in which the intended conductive pattern is opaque to light, and the areas intended to be free from conductive material are transparent.

Positive-acting resist. A resist which is decomposed (softened) by light and which, after exposure and development, is removed from those areas which were under the transparent parts of a production master.

Printed board. The general term for completely processed printed circuit or printed wiring configurations. It includes rigid or flexible, single, double, and multilayer boards.

Printed circuit. A conductive pattern comprised of printed components, printed wiring, or a combination thereof, all formed in a predetermined design and intended to be attached to a common base. (In addition, this is a generic term used to describe a printed board produced by any of a number of techniques.)

Printed component. A part, such as an inductor, resistor, capacitor, or transmission line, which is formed as part of the conductive pattern of the printed board.

Printed contact. A portion of a conductive pattern formed by printing, serving as one part of a contact system.

Printed wiring. The conductive pattern intended to be formed on a common base, to provide point-to-point connection of discrete components, but not to contain printed components.

Printed wiring assembly drawing. A document that shows the printed board (rigid or flexible), the separately manufactured components which are to be added to the board, and any other information necessary to describe the joining of these parts to perform a specific function.

Printed wiring layout. A sketch that depicts the printed wiring substrate, the physical size and location of electronic and mechanical components, and the routing of conductors that electrically interconnect components, in sufficient detail to allow the preparation of documentation and artwork. (See Fig. 2.)

Production master. A 1 to 1 scale pattern which is used to produce one or more printed boards (rigid or flexible) within the accuracy specified on the Master Drawing. (See Fig. 2.) *Single-image production master.* A production master used in the process of making a single printed board. *Multiple-image production master.* A production master used in the process of making two or more printed board simultaneously.

Pull strength.* See Bond strength.

R

Reference edge. The edge of cable or conductor from which measurements are made. (Sometimes indicated by a thread, identification stripe, or printing. Conductors are usually identified by their sequential position from the reference edge, with number one conductor closest to this edge.)

Reflow soldering. A process for joining parts by tinning the mating surfaces, placing them together, heating until the solder fuses, and allowing to cool in the joined position.

Register mark. A symbol used as a reference point to maintain registration.

Registration. The degree of conformity of the position of a pattern, or a portion thereof, with its intended position or with that of any other conductor layer of a board.

Repairing. The act of restoring the functional capability of a defective part without necessarily restoring appearance, interchangeability, and uniformity.

Resist. Coating material used to mask or to protect selected areas of a pattern from the action of an etchant, solder, or plating.

Resistance soldering. A method of soldering in which a current is passed through and heats the soldering area by contact with one or more electrodes.

Reverse Image. The resist pattern on a printed board used to allow for the exposure of conductive areas for subsequent plating.

Reworking. The act of repeating one or more manufacturing operations for the purpose of improving the yield of acceptable parts.

Ribbon cable.* See **Flat Cable**.

---

*The preferred term is referenced.

Right-angle edge connector. A connector which terminates conductors at the edge of a printed board, while bringing the terminations out at right angles to the plane of the board conductors.

Roadmap. A printed pattern of non-conductive material by which the circuitry and components are delineated on a board to aid in service and repair of the board.

Robber. * See **Thief.**

S

Schematic diagram. A drawing which shows, by means of graphic symbols, the electrical connections, components, and functions of a specific circuit arrangement.

Screen Printing (nonpreferred synonym: Silk screening). A process for transferring an image to a surface by forcing suitable media through a stencil screen with a squeegee.

Seeding. * See **Activating.**

Seed layer. * See **Activating.**

Semiadditive process. An additive process for obtaining conductive patterns which combines an electroless metal deposition on an unclad substrate with electroplating with etching, or with both.

Sensitizing. * See **Activating.**

Shielding, electronic. A physical barrier, usually electrically conductive, designed to reduce the interaction of electric or magnetic fields upon devices, circuits, or portions of circuits.

Signal. An electrical impulse of a predetermined voltage, current, polarity, and pulse width.

Signal conductor. An individual conductor used to transmit an impressed signal.

---

*The preferred term is referenced.

Signal plane. A conductor layer intended to carry signals, rather than serve as a ground or other fixed voltage function.

Silk screen (verb).* See **Screen printing.**

Single-image production master.* See **Production master.**

Single-sided board. A printed board with a conductive pattern on one side only.

Smear removal.* See **Chemical hole cleaning.**

Solder mask coating.* See **Resist.**

Solder plugs. Cores of solder in the plated-through holes of a printed board.

Solder projection. An undesirable protrusion of solder from a solidified solder joint or coating.

Solder resist.* See **Resist.**

Solder side. The side of a printed board which is opposite to the component side.

Solderability. The property of a metal to be wetted by solder.

Soldering. A process of joining metallic surfaces with solder, without the melting of the base material.

Span. The distance from the reference edge of the first conductor to the reference edge of the last conductor, expressed in decimal inches or centimeters.

Spurious signal.* See **Crosstalk.**

Step-and-repeat. A method by which successive exposures of a single image are made to produce a multiple-image production master.

Stripline. A type of transmission line configuration which consists of a single narrow conductor parallel and equidistant to two parallel ground planes.

---

*The preferred term is referenced.

Substrate.* See Base material.

Subtractive process. A process for obtaining conductive patterns by the selective removal of unwanted portions of a conductive foil.

T

Tab.* See Printed Contact.

Taped components. Components attached to a continuous tape for and in automatic assembly.

Tenting. A printed board fabrication method of covering over plated-through holes and the surrounding conductive pattern with a resist, usually dry film.

Terminal area.* See Land.

Terminal clearance hole.* See Access hole.

Terminal hole.* See Component hole.

Terminal pad.* See Land.

Test coupon. A portion of a printed board or of a panel containing printed coupons, used to determine the acceptability of such a board(s).

Test pattern. A pattern used for inspection or testing purposes.

Test point. Special points of access to an electrical circuit, used for testing purposes.

Thermal relief. Crosshatching to minimize blistering or warping during soldering operations.

Thin foil. A metal sheet less than 0.007 in. (1/2 oz) thick.

Through connection. An electrical connection between conductive patterns on opposite sides of an insulating base, e.g., plated-through hole or clinched jumper wire.

----

*The preferred term is referenced.

Tie Bar.* See **Plating bar.**

Tinning. A process for the application of solder coating on component leads, conductors and terminals, to enhance solderability.

Tooling feature. A specified physical feature on a printed board, or a panel, such as a marking, hole, cut-out, notch, slot or edge, used exclusively to position the board or panel or to mount components accurately.

Tooling holes. The general term for holes placed on a printed board, or a panel, and used to aid in the manufacturing process.

Transmission cable. Two or more transmission lines. (If the structure is flat, it is called "flat transmission cable" to differentiate it from a round structure, such as a jacketed group of coaxial cables.) (See also Transmission line.)

Transmission line. A signal-carrying circuit composed of conductors and dielectric material with controlled electrical characteristics used for the transmission of high-frequency or narrow-pulse type signals.

Trim lines. Lines which define the borders of a printed board. (See also Corner mark.)

True position. The theoretically exact location of a feature or hole established by basic dimension.

True position tolerance. The total diameter of permissible movement around the true position as shown in the master drawing.

Twist. The deformation parallel to a diagonal of a rectangular sheet such that one of the corners is not in the plane containing the other three corners.

Two-sided board.* See **Double-sided board.**

---

*The preferred term is referenced.

U

Undercut (in process). The distance on one edge of a conductor measured parallel to the board surface from the outer edge of the conductor, including etch resists, to the maximum point of indentation on the copper edge. (See Fig. 10.)

Undercut (after fabrication). The distance on one edge of a conductor measured parallel to the board surface from the outer edge of the conductor, excluding overplating and coatings, to the maximum point of indentation on the same edge. (See Fig. 10.)

Underwriters symbol. A logotype authorized for placement on a product which has been recognized (accepted) by Underwriters Laboratories, Inc. (UL).

V

Via hole. A plated-through hole used as a through connection, but in which there is no intention to insert a component lead or other reinforcing material.

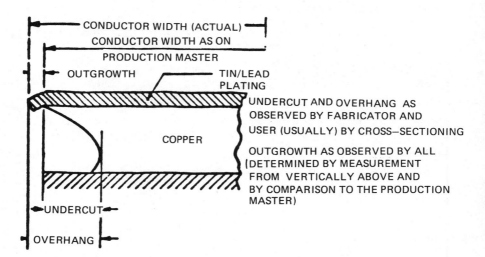

Figure 10. Undercut.

Void. The absence of substances in a localized area.

Voltage plane. A conductor or portion of a conductor layer on or in a printed board which is maintained at other than ground potential. It can also be used as a common voltage source, for heat sinking, or for shielding.

Voltage-plane clearance. Voltage-plane clearance is the etched portion of a voltage plane around a plated-through or nonplated through hole that isolates the voltage plane from the hole. (See Fig. 6).

W

Warp.* See Bow.

Wave soldering. A process wherein printed boards are brought in contact with the surface of continuously flowing and circulating solder.

Weave exposure. A surface condition of base material in which the unbroken fibers of woven glass cloth are not completely covered by resin.

Weave texture. A surface condition of base material in which a wave pattern of glass cloth is apparent although the unbroken fibers of the woven cloth are completely covered with resin.

Wetting. The formation of a relatively uniform, smooth, unbroken, and adherent film of solder to a base material.

Whisker. A slender acicular (needle-shaped) metallic growth on a printed board.

Wicking. Capillary absorption of liquid along the fibers of the base material.

---

*The preferred term is referenced.

Index